To
Professor Leon Lederman
With Best Wishes
Manjit Kumar

SCIENCE AND THE RETREAT FROM REASON

Science and the Retreat from Reason

JOHN GILLOTT
MANJIT KUMAR

MERLIN PRESS
LONDON

First published in 1995
by The Merlin Press Ltd
10 Malden Road
London NW5 3HR

© The Merlin Press Ltd

ISBN 085036 451 5 (Hardback)
ISBN 085036 433 7 (Paperback)

Typesetting by Computerset
Harmondsworth, Middlesex
Printed in Finland by WSOY

For Josephine and Pandora

Contents

Extended Contents

CHAPTER FIVE: SCIENCE AND HUMANISM

CHAPTER SIX: SCIENCE AND THE RETREAT FROM REASON

CHAPTER SEVEN: THE LOSS OF CERTAINTY AND THE QUEST FOR BEAUTY IN SCIENCE

Denigrate Humanity • Over the Top against a Theory of Everything • Roger Penrose, Superstrings and the Limits to Speculation • When Speculation Masquerades as Theory • Lessons from Kepler • The Flight into Mathematics • Empirical Results Temporarily Postpone Operations • Steven Weinberg, Simplicity and Inevitability • The Quest for Beauty as a Problem for Science • Moving the Goalposts in Particle Physics • Science and Art • Mathematics and the Loss of Certainty • Reversing the Retreat from Reason

NOTES AND REFERENCES

BIBLIOGRAPHY

INDEX

Acknowledgements

In the course of writing this book, we benefited from the suggestions and criticisms of a range of people with interests in science, history, and politics. We would like to express our thanks to Chris Isham, for comments on two versions of Chapter Three; to John Maynard Smith, for comments and suggestions on Chapter Four; to Pandora Kay-Kreizman, for her comments and suggestions on the book as it developed; and to Michael Fitzpatrick, for his incisive suggestions on structure and argument, as well as his constant enthusiasm for the book as it worked its way to completion.

Particular thanks go to our editor, James Woudhuysen, of the Henley Centre for Forecasting. James insisted that we made each of the issues we dealt with as accessible as possible to a lay readership – by spelling out enough detail to make our arguments convincing. He also suggested useful ideas, especially for Chapter Six. Finally, over several iterations, he put in many evenings and weekends making our prose readable. Thank you, James.

Finally, we would like to thank Julie Millard and Martin Eve at Merlin Press for their support and their patience.

Of course, as is always the case in these matters, we take responsibility for the arguments in this book.

John Gillott and Manjit Kumar, London, June 1995

Introduction:
The Sense of an End

In 1977, Robert Edwards and Patrick Steptoe fertilized a human egg outside a woman's body. They then successfully transferred it into the mother's uterus, after which the pregnancy proceeded as had many billions before. This achievement, known as *in vitro* fertilization (IVF), led to the birth of Louise Brown, who was promptly dubbed the world's first 'test-tube baby'.

Seven years later, America launched the multi-billion dollar Human Genome Project. The objective was to produce a comprehensive map of the basic molecular codes of human inheritance. The Project also aims to work out the precise location of these genetic codes on human chromosomes.

As we approach the new millennium, further advances in treatments for infertility, beyond those developed by Edwards and Steptoe, are on offer. For the estimated 10 per cent of couples in the West who are infertile, there is hope. Worldwide, the promise which the Human Genome Project holds is enormous: that of preventing – and perhaps even treating – inherited diseases, from Huntington's disease to cystic fibrosis.

In practice, the life sciences are moving ahead. Yet the dominant popular response to dramatic leaps forward in fertility treatments and genetics is apprehensive. The media focus on the possible dangers of the new techniques. They point out the scope for their abuse. Treatments for infertility amount, as far as a great deal of press coverage goes, only to the 'yuk factor' which allegedly surrounds the donation of eggs and artificial insemination.

The insights provided by genetic engineering today tend, often, to prompt lurid tales of Dr Josef Mengele and his grisly experiments in the Nazi concentration camps. Few discuss the potential genetic engineering has for ending Tay-Sachs disease, thalassaemia, or muscular dystrophy. The general preoccupation of the media and of officialdom is with the need to curb the new techniques. Rather than improve our understanding of biology and make practical innovations in the field more widely available, a consensus has emerged that scientists are going too far and too fast. There was outcry when, in 1994, a 59-year-old British woman travelled to see a specialist in Italy and had a baby. Scientists' intellectual arrogance and interventionist zeal, it was held, would ensure that such projects ended in tears.

This fatalistic attitude towards scientific advance now permeates popular culture. Stephen Spielberg's film of Michael Crichton's novel *Jurassic Park* (1993) indicted genetic engineering. In 1994, Kenneth Branagh's remake of Mary Shelley's *Frankenstein* offered yet another version of the classic tale of the evils unleashed by scientific meddling in the mysteries of nature. But mass-audience excursions on the menace of science do not just revolve around the animal kingdom. A comprehensive survey of films about the future made in the West in the 1970s and early 1980s reveal that only three among 52 showed 'anything resembling the triumph of progressive technology.'[1] Indeed the whole idea of technology working to the advantage of humanity has been discredited. As the cultural critic Andrew Ross has acutely remarked, the period 'look' of the future 'is a survivalist one, governed by the dark imagination of technological dystopias.'[2]

Losing Faith in the Future

Today, it is true, there is a large readership for popular books about science. However, such works often tend toward mysticism and metaphysics, the historic enemies of scientific thought. Paul Davies' books on the scientific basis for religion – including *God and the Big Bang*, and *The Mind of God* – are characteristic of a strong mystical tendency in popular science writing. Clerics have given an enthusiastic welcome to this trend. Don Cupitt has observed, under the headline 'Onward Crypto Theists', that not only physicists, but also the general public are gratified by the idea that theoretical physics is really 'a branch of theology.'[3]

If science, in the mass-market bookshop, has blurred into religion, prejudice can all too easily flourish. In the 1990s, after all, quack remedies have become popular in fields as varied as medicine, mental health and management science. Given such a climate of superstition, it is not surprising to find a sense of despondency among scientists today.

Nobel Prize winner Leon Lederman, one of America's most distinguished scientists, has raised the right questions. In his provocatively-titled research survey *Science: The End of the Frontier?* (1991),[4] he contrasts today's 'mood of uncertainty and discouragement' about science with the hopes of the post-war years. 'Once upon a time', Lederman notes, American science 'sheltered an Einstein, went to the moon, and gave to the world the laser, the electronic computer, nylon, television, the cure for polio, and observation of our planet's position in an expanding universe.'[5] Now, however, things look very different.

The research conducted for *End of the Frontier?* was interesting for everyone concerned with the current mood of the scientific community. In his capacity as the then president-elect of the American Association for the Advancement of Science, Lederman arranged for America's leading scientists to be sent self-completion questionnaires on their feelings about contemporary science and its organisation. The results revealed widespread gloom:

> The response paints a picture of an academic research community beset by flagging morale, diminishing expectations, and constricting horizons. From one institution to the next, across demographic categories, across disciplines of research, the nation's scientists are sending a warning. Academic research in the United States is in serious trouble.[6]

Lederman suggests that the common theme underlying public and government attitudes toward science is 'a loss of faith in the future'.

What a contrast with the bold optimism of the scientists of the past! In the seventeenth and eighteenth centuries, the pioneers of modern mathematics, physics, chemistry and biology were imbued with a sense of the boundless possibilities science offered in the service of humanity. Back then, when economic and social developments often outpaced the spread of popular demands for liberty and democracy, science was at the cutting edge of philosophical and cultural advance. Then, the growth of scientific ideas undermined ecclesiastical authority and raised the prestige of science.

Despite two world wars and distinct bouts of depression about its prospects, science and confidence about it, survived intact for much of the twentieth century. The spirit of inquiry and experimentation which had once animated the science of the Enlightenment continued, albeit in a manner which suited the Cold War. There was a tendency for pure science to be subordinated to technology, and for the imperatives of the military sector to take priority over civilian needs. However, as Lederman and many others have noted, significant advances were made in many areas, including nuclear power, the exploration of space, and computers.

Science also continued to play a prominent public role in the life of society. From Franklin D. Roosevelt's drive to harness science to US national reconstruction in the 1940s through to Harold Wilson's invoking of the 'white heat of technological revolution' in Britain in the 1960s, science retained a high political profile, and scientists were held in high regard.

Today, however, attitudes toward science are much more ambiguous. Over the past 30 years, the conviction has grown that scientific activity is more likely to result in harm than good. The growing consensus behind 'Green' ideas, which broadly elevate the natural world above human attempts to modify it, has put science on the defensive.

Until the late 1960s, science was generally regarded as laying the basis for progressive interventions in a natural world viewed as threatening, capricious, and potentially destructive. But in the wake of DDT, Bhopal and Chernobyl, it is science that has come to be seen as a malign force. Moreover, if science's interference with nature is seen as self-serving, nature is now perceived as essentially benign and harmonious.'

Pure Research in Check

The decline of state-sponsored research and development parallels the fall of science from public grace. In the past, government programmes to develop atomic weapons and conquer space encouraged both theoretical enquiry, in the hope of long-term breakthroughs, and work on the practical applications of recent discoveries. But in the past decade that picture has changed. Pressures to curb government expenditure have led the authorities to place a premium on technological spin-offs offering short-term commercial advantage. 'Pure' research, as opposed to the 'applied' sort, brings little immediate pay-off. It is now held in check by governments right across the globe.

In 1993, the US Congress decided to cancel America's Superconducting Super Collider (SSC), a facility in Texas for basic research into the structure of matter. In effect, the move signalled the end of nearly half a century of backing for pure research. Back in 1945, Vannevar Bush, head of the US Office of Scientific Research and Development, published a report entitled *Science, the Endless Frontier.* In it he envisaged endless benefits to American society from scientific advance. Science would bring jobs, rising living standards, and improvements in culture. Just as importantly, Bush also emphasised the significance of pure science. 'Basic research', he contended, 'leads to new knowledge'. And this was a vital resource for the future of society because

it provides scientific capital. It creates the fund from which the practical applications of knowledge must be drawn.[8]

Some 48 years after Bush, the abandonment of the Super Collider cost thousands of jobs. It also severely undermined what, by any standards, remains a significant branch of research. In the past, theoretical research as 'pure' as that which demanded a Super Collider led, albeit indirectly and over decades, to microelectronics and to lasers. The fate of the Collider shows that the luminous platform for scientific effort outlined by Bush has in practice collapsed.

Anti-Science Symptoms Reflect Political and Social Causes

The morale of the scientific community has declined. In the 1950s and 1960s, top scientists joined the establishment and acquired celebrity status on both sides of the Atlantic. But today's scientists occupy a more insecure position. There is still a scattering of science celebrities; but, when they do appear in public, it is more often than not to plead greater recognition and funding for science. Or else it is to defend their various projects from resource cuts and – an equal menace, in some fields – ideologically-motivated attacks.

On both sides of the Atlantic, scientists have become preoccupied with what they detect as distinct and novel *anti-science trends* in society. In response, they have set up high-level committees, sponsored lobby groups, and published special journals – all to promote the cause of science to politicians, business figures and the general public. Yet the odd thing is that such frenetic moves amount more to panic than to anything else.

Despite public fears about some aspects of scientific research, and despite vague but discernable public suspicion of scientists, a thoroughgoing hostility to science remains exceptional in society today. Everyone has become much too dependent on new technologies, and too familiar with scientific novelties, for damning repudiations of science to become widely influential. Projects in theoretical physics may be thrown into jeopardy for economic reasons; but biology and computer science, to name but two branches of contemporary enquiry, are thriving.

Anti-science *prejudices* in society are real enough. But they are neither new nor, by themselves, a major threat to scientific advance. The real problem lies in the *changed relationship between science and society*. We believe that it is this that underlies the spread of anti-scientific attitudes today.

At the centre of the scientific optimism of the past lay the conviction that science was but one tool in a broader project of human social advance. Just as the philosophers of the Enlightenment held that the power of reason could improve society, so their colleagues in the natural sciences believed that research and experiment could benefit the human condition. Science marched in the vanguard of *progress*. The definition given this word by a typical nineteenth-century French dictionary is worth recording here:

> Humanity is perfectible and it moves incessantly from less good to better, from ignorance to science, from barbarism to civilisation... The idea that humanity becomes day by day better and happier is particularly dear to our century; faith in the law of progress is the true faith of our century.[9]

The 'perfectibility' of human beings, and thus the project of progress through rational enquiry, was first put forward by the mathematician, philosopher, and member of France's National Assembly, the Marquis de Condorcet [1743–1794].

The Marquis and his *Picture of a Historical Tableau of the Progress of the Human Mind* (1794) represented one camp. But they came to be roundly attacked by an English parson, the Reverend Thomas Malthus [1766–1834], in his famous *Essay on the Principle of Population* (1798). Nevertheless, for years Condorcet's project served as a stated or unstated principle informing the rise not just of science, but of modern civilization.

The situation now is very different. Better lives for happier people through better science: this seems an unlikely prospect, if not an impossible dream. Contemporary disillusionment with science has not cohered into a fresh and comprehensively black anti-science trend. But the growth in prejudices and hostile attitudes toward particular branches of science is real enough. And it is real because it is a symptom of a broader rejection, by society, of the project of progress.

Behind the Crisis in Science

Modern thought betrays a deep loss of confidence in the possibility of constructive intervention in either the social order or the natural one. Such a pessimistic outlook has been articulated in the past; but, this time, mass psychology seems more hopeless about prospects than at any time before.

At the close of the nineteenth century, the industrial and commercial expansion of the West was undeniably spectacular. Yet the West soon failed to deliver on its pledge of a free, democratic and civilised society. Two world wars, the Depression, fascism, the Holocaust, Hiroshima, the Cold War: the story of much of the past 100 years has been a barbaric one. The cumulative impact of all these events has fostered a mood of deep anxiety about the future. The problems of modern society, many feel, are now intractable.

There is, too, a ready audience for those who blame such problems not on the prevailing forms of social and economic organisation, but on science, and indeed on rational enquiry and analysis. As Václav Havel, President of the Czech Republic, explained what he called 'the end of the modern era' in 1992:

> The fall of communism can be regarded as a sign that modern thought – based on the premise that the world is objectively knowable, and that the knowledge so obtained can be absolutely generalised – has come to a final crisis...
>
> Modern rationalism and modern science, through the work of man that, as all human works, developed within our natural world, now systematically leave it behind, deny it, degrade and defame it – and, of course, at the same time, colonize it.[10]

Altogether, pessimism about society's prospects has driven

people to a more general attack on reason and science.

Of course, conservative philosophers and politicians have denounced these things ever since the French Revolution. What is new today is that the anti-modernist case has also been adopted by many who used to profess liberal, pro-science, views. In America, for example, George E. Brown was for many years chair of the Congressional Committee on Science, Space and Technology. He was a staunch defender of science. In an article published in 1992, however, Brown suggested a reconsideration of the relationship between government and science, observing that 'the uneasy alliance between scientists and politicians' was beginning to come 'unglued'.[11] And in 'The Objectivity Crisis', another statement made the same year, he was blunter still. 'The promise of science', he argued, might be 'at the root of our problems.'[12]

The relative strength of anti-progress thinking is also revealed by the strikingly defensive posture of those scientists who do still seek to uphold traditions of experimentation and social engagement. Thus, in a riposte to critics of science, the distinguished American historian of science Gerald Holton can only offer, as an alternative, a rearguard campaign to challenge superstition in US schools. This narrow response is unlikely to have much effect since, as Holton himself notes, the more educated people are at present, the more likely they are to be hostile toward science.[13] In a similar way Peter Scott, former editor of Britain's *Times Higher Education Supplement*, upholds reason and science, but makes concessions all the same: knowledge, he writes, must now embrace 'the affective, the indeterminate, the unsure.'[14]

Behind the current crisis of science, we have ventured, lies the collapse of confidence in progress. The very idea of trying to improve human society is now held to be at best misguided, at worst dangerous. The emphasis of the Green movement on the natural *limits* to human activity in all spheres – in economics and in population especially – acts as a check on human goals. Today's rejections of science, in short, are founded on a lack of faith in humanity.

The central aim of this book is to put afresh the argument for human progress. Such progress, we believe, will partly depend on intervention in nature. We do not, however, simply reassert past visions of progress. Reflecting the constraints of their times, many such visions have had a mechanistic character about them: they have upheld not science, but what we prefer to term *scientism*.

Our conception of progress is different. We believe that progress results from human actions – actions which throw up problems at the same time as they contribute solutions. Indeed, it is only through the creative process of problem-solving and identifying new problems that the dynamic of progress unfolds.

The book is divided into two halves. Chapters Two, Three and Four critically examine the claim that developments in twentieth-century science invalidate any project which seeks to use mastery over nature as the basis for progress. In these three chapters we focus on quantum mechanics, chaos theory and complexity theory. Our conclusion is that there is no basis in natural science for such bleak claims.

Chapters Five through to Seven present a sociological and historical explanation of the breakdown of modern conceptions of progress. Chapter Six – *Science and the Retreat from Reason* – has the same title as the book. It takes the disjunction between continuing scientific advance and the deepening unpopularity of progress as its theme. It shows how this disjunction has come about, and explores its damaging consequences for both society and science.

We begin by setting the scene in Chapter One. Here we open with a paradox: that the historic boundary against which science seems now to be pressed up was in fact given in the nature of Vannevar Bush's 'endless frontier'.

Chapter One:
The Post-War Loss of Certainty

In his *Science and Religion: Some Historical Perspectives* (1991), the historian John Hedley Brooke makes an important point. He believes that it is the twentieth century as a whole, and not just the recent era, which is characterised by a collapse of scientific optimism:

> The twentieth century has had its crisis of faith, with the loss of that confidence that in the Enlightenment had been placed in science as the key to solving all human problems.[1]

Brooke's periodisation of scientific *anomie* seems to us significant, in that it would include the sunlit years after 1945. Then attitudes toward science were forward-looking by comparison with today's insecurities. But they were also attitudes framed by a philosophy of progress which, as we show in this chapter, was at bottom misanthropic.

In both duration and substance, Bush's 'endless frontier' compared poorly with the stir caused by the Enlightenment of the seventeenth and especially the eighteenth centuries. Enlightenment thinkers enthused about science because they approved of progress; Bush and others, by contrast, endorsed science in their urge to prosecute the Cold War. They favoured progress, but only in the misanthropic sense of increasing American military power.

From Bacon to Condorcet

In the early seventeenth century, learned discussion came to a consensus on three core themes. First, leading thinkers began to have faith in human abilities and in human judgement. Second, opinion turned to favour the idea that the manipulation of nature would improve humanity's lot. Finally, the feeling grew that if human faculties were admirable, and if the manipulation of nature could improve human society without limit, then the world could look forward to humanity realising its potential more and more. Human *perfection* might prove unreachable; but the *perfectibility* of the species, in the sense of its capacity always to raise itself to new heights – this, even before Condorcet, was fundamental to the Enlightenment world-view.

Francis Bacon [1561–1626] was the pioneer of the modern scientific method. He also wrote widely on the three themes we have outlined.

Entering Parliament at the age of 23, Bacon became Lord Chancellor in 1618. Two years later, he was forced to leave public life after admitting to having taken a bribe. Yet Bacon's loss proved to be posterity's gain. Adding significantly to the body of writing he had already produced, Bacon went on to dedicate his last years to study, and a still more prolific output.

'Knowledge is Power', Bacon's famous dictum, is often taken as a summation of his view of the importance of science. Its aim, he maintained, should not just be 'the knowledge of causes, and secret motions of things', but also 'the enlarging of the bounds of human empire, to the effecting of all things possible.'[2] Much of Bacon's pre-eminence consists in the fact that he was one of the first thinkers to reject any limits to human achievement.

René Descartes [1596–1650], one of the founders of modern philosophy, differed with Bacon on the *origins* of knowledge. Bacon emphasised learning from experience: in essence, his was the method of induction. Descartes, by contrast, held that knowledge began from certain truths, from which all else could be derived. Broadly, he favoured the method of deduction.

What united both Bacon and Descartes, however, was a similar notion of the purpose of knowledge. In a striking passage in his *Discourse on Method* (1637), Descartes argued that science should help humanity control nature. Attacking the 'speculative philosophy taught in schools', he extolled, instead, 'a practical philosophy' which knew 'the power and the effect of fire, water, air, the stars, the heavens and all the other bodies which surround us'; a philosophy which knew things as intimately as craftsmen knew their trades.

Such a philosophy, Descartes continued, might, like a craftsman, put insights 'to all those uses for which they are appropriate'. For Descartes, then, the purpose of knowledge was clear. It was to 'make ourselves, as it were, masters and possessors of nature.'[3]

The optimistic perspective of Bacon and Descartes gained adherents in England before it was disseminated elsewhere. The Royal Society helped here. Founded in 1660, the Society's specific role was to publicise the 'new philosophy', as the emerging consensus about knowledge, science, and their role came to be known.

Growing respect for humanity's intellectual and practical conquests was accompanied by declining reverence for the older traditions of science developed in ancient Greece. The historian Richard Westfall conveys the process well in his *The Construction of Modern Science*, where he points to Joseph Glanvill's *Plus Ultra* (1688). That book, Westfall argues,

relied primarily upon science to argue that modern achievements had surpassed the ancients. The title itself implies the content. A conscious play on the ancient myth that the pillars of Hercules at the straits of Gibraltar bore the motto *'ne plus ultra'* (go no further), Glanvill's title, *Further Yet*, proclaimed that the narrow limits of the ancient intellectual world had to be torn asunder. *Plus Ultra* is a catalogue of modern achievements, primarily scientific discoveries. Glanvill listed accomplishments in anatomy, optics, and chemistry. He cited inventions – the microscope, telescope, barometer, thermometer, air pump. The whole tone of the work expresses the fact that authority did not command his allegiance.[4]

Glanvill typified a fresh, open-ended willingness to confront new challenges – an outlook which later cast its spell over a generation of English scientists.

Humphry Davy [1778–1829], later in life the director of the Royal Institution of Great Britain, was one of those scientists. By the beginning of the nineteenth century, science had, Davy contended, bestowed upon humanity 'powers which may be called creative': powers which enabled humanity to 'interrogate nature with power'. For Davy, science helped humanity enquire into nature not 'simply as a scholar, passive and seeking only to understand her operations, but rather as a master.'[5]

In the seventeenth and early eighteenth centuries, English scientists such as Davy did much to popularise the idea of humanity as Emperor of the Natural World. It was the thinkers of the succeeding French Enlightenment, however, who most explicitly developed a vision of progress that went beyond scientific advance. In their vision, humanity was no longer a static species: it was capable of endless development – through increasing its powers over nature.

Inspired by the achievements and promise of the French Revolution, Condorcet was, as we noted, a champion of human development. He wrote:

No limit has been set to the improvement of human faculties, that the perfectibility of man is really boundless, that the progress of this perfectibility, henceforth independent of any power that would arrest it, has no other limit than the duration of the globe where nature has set us. No doubt this progress may be more or less rapid, but there will never be any retrogression.[6]

In 1794, Condorcet felt that the curve of human development could only be upward: 'there will never be any retrogression.'

Today, by contrast, the possibility of broad social advance needs upholding.

Why? One reason is that the 'retrogression' which Condorcet ruled out has certainly occurred in the twentieth century. Another reason lies in reactions to the century's catastrophic and bloody events. These events, from the Boxer uprising in China in 1900 to the bombing of Oklahoma City in 1995, imply, for a generation of Western intellectuals, not that social advance must be fought for, but that it, like science, ushers in too many problems to be a cause worth fighting for.

Twentieth-Century Blues: Truth Ruled Out

In his *History of Western Philosophy*, written during the Second World War, Bertrand Russell pinpointed the opening year of the First World War as the end of a period of progress stretching back nearly a millennium:

> The year 1000 may be conveniently taken as marking the end of the lowest depth to which the civilization of Western Europe sank. From this point the upward movement began which continued till 1914.[7]

After the First World War was over, a kind of mental paralysis did indeed grip European élites and, to an extent, popular thinking all over the industrialised world. The war, after all, had shown that the metropolitan nations had deep conflicts of interest. Economic and social dislocation during and after the war also ran deep. The success of the Bolsheviks in Russia seemed to confirm many people's worst fears. Before the war, forthright opponents of scientific and social progress were in a minority. After it, books such as Oswald Spengler's *The Decline of the West* (1918) reached a mass audience.

After 1945, dark moods appeared to lift. But Bush's 'endless frontier' for science was a qualified enterprise from its inception. The post-war reception of Karl Popper's ideas, shows that Western theorists recoiled from the idea of bending nature, through science, to human will.

In *The Logic of Scientific Discovery* (1934), his famous attack on the method of induction, Popper dismissed the idea that scientific knowledge develops when people generalise from experience and observation. From there, he argued that a scientific theory can never be proved to be true: it can only be proved false. This, in substance, was Popper's famous theory of 'falsification'. Its popularity at the time of Bush shows, in our view, the irrational currents which provided the real undertow to the Bushian vision. To explain this requires that we briefly first clarify some basic issues of logic.

Inductivism, Hume and Russell

Logic is the study of reasoning, of arguments bridging premises to a conclusion. The aim is to state, in as general terms as possible, the conditions under which an argument is valid, and to identify structures or forms of valid argument.

Deductive logic is concerned with the special case when reasoning is conclusive. Given that the premises are true, the conclusion *must* be true. There is absolutely no possibility that the premises are true, and the conclusion false. The premises guarantee the truth of the conclusion.

Most everyday human reasoning, however, is in fact very far from deductive. Consider the following conversation:

'The buffet car is at the front of the train.'
'How do you know?'
'I've seen lots of people coming from that direction carrying drinks and sandwiches.'

The final statement appears to be sensible reasoning; but it cannot be deduced from the premise. After all, the passengers spotted by one of the conversants may have overshot their seats on the train – they may come from the direction they do because they are retracing their steps from a buffet at the back of the train, not at the front.

Deductive logic does not get very far in reasoning about matters of fact. It merely draws out what is implicit in the premises. No risk is involved, and no extension of knowledge is made. Inductive logic, by contrast, is the study of non-conclusive reasoning, of the means by which we feel sure that, despite the incompleteness of the evidence before us, we are nevertheless confident that the buffet car really is at the front of the train. Inductive logic is of immense importance not only in normal life, but also in science. *Inductivism*, the philosophy based on it, argues that science induces its laws and theories from empirical facts. However, inductivism faces a number of difficulties.

One such difficulty was outlined by the Scottish philosopher David Hume [1711–76] in his *Treatise on Human Nature*. Referring to what he described as 'the problem of induction', Hume asked: What reason can anyone have for assuming that future observations would resemble past ones? For example, why should the sun rise tomorrow morning – apart from the fact that it has always done so in the past? Consequently, on what grounds should we ever accept, uncritically, those 'laws of nature' which are merely induced from observed facts?

Beyond philosophical circles, Hume's critique of inductivism was effectively dissolved by the success of the scientific enterprise in the nineteenth century. The ubiquity of Popper's ever more agnostic critique in the era of Bush, however, showed that, by then, conceptions of science had changed enormously. Enlightenment certitudes had been lost. The whole relevance of the scientific project was in grave doubt.

Bertrand Russell waged a fruitless rearguard campaign on behalf of inductivism. In an attempt to counter Hume's objection, he formulated a 'principle of induction'. Russell's argument was that the more often a phenomenon was observed side by side with another, the more likely it was that they were causally related in some way. Translated into a theory of the link between scientific theories and observations, this idea became: the more often a theory accorded with observations, the more likely it was to be true. Truth was, in this sense, achievable.

Karl Popper and the Logic of Scientific Irrationalism

Popper rejected Russell's principle. First, with regard to particular correlations, he asked: how can we say all swans are white just because we have not seen any black ones? It could be the case that the next swan seen is black. No amount of observing white swans allows any inferences to be made about the probability of the next swan being white. Second, Popper pointed out that, though Newtonian theory had very often accorded with observations, Einstein's general theory of relativity had eventually shown that it was fundamentally wrong. For Popper this confirmed that inductivism had inherent defects.

Our riposte to Popper is that human beings nearly always assume that nature has a definite character. That character, they try to understand through theories, which they arrive at through a combination of guesswork and induction. Each theory is better than the last – to the extent, that is, that each captures a further aspect of the truth.

Each new theory is however also false, for it is not a total theory. Thus Newton's theory of gravity is both approximately true, and also false. No doubt the same is true of Einstein's theory of general relativity. As for the world's swans, we anticipate whiteness because we generalise from repeated sightings of white swans, and estimate that whiteness will be the rule because of what we know of evolution and genetics.

We may later revise our theory of what is the cause of a particular swan's whiteness. But, assuming that there is an objective character to natural law – which Popper was inclined to do – our move from the particular to the general is justified.

But it is just this assumption of an *objective reality that human beings can understand* that Popper opposed. For him, scientific discovery and the genesis of ideas lay beyond comprehension. 'Every discovery', Popper wrote, 'contains "an irrational element", or "a creative intuition" in Bergson's sense.'[8]

In his excellent *The Rationality of Science*, W. H. Newton-Smith puts his finger on Popper's error. To abandon induction, as Popper did, is to undermine 'the claim that there is a growth of scientific knowledge and that science is a rational activity.'[9] Popper wanted to have it both ways: he characterised science as the rational pursuit of truth, but at the same time he held that 'the possession of truth is not recognisable.'[10]

Newton-Smith's casting of Popper as an 'irrational rationalist' is very apt. But if science does not originate simply in intuition, we accept that intuition plays an important role in the development of scientific theories. However, the flashes of brilliance upon which scientific theories partly depend occur in the context of the accumulated knowledge of the scientific community as a whole. Discoveries emerge out of – and rise above – the intellectual and cultural climate and metaphors of their time. Popper confused the individual act of discovery, which will always contain personal and therefore subjective elements, with the broader prerequisites for the development and progress of scientific knowledge.

For Popper, natural science could not aspire to Final Truth. From this, Popper went on to argue that similar aspirations for human conduct were equally misguided. All that was on the agenda, for society, he wrote in his *Poverty of Historicism* (1944), was the work of the 'piecemeal social engineer'.[11]

In effect, Popper condemned any kind of root-and-branch change as witchcraft. A 'holistic' approach to society was, for him, 'pre-scientific'. For Popper, radical social change contained 'an element of perfectionism', and was therefore to be derided. He went on:

> Once we realise... that we cannot make heaven on earth but can only improve matters a little, we also realise that we can only improve them little by little.[12]

The emphasis in Popper was always on conserving the past, on gradualism. In his opinion, it would be wrong to expect too much, either of science or of society. The British historian E. H.

Carr once made this observation:

> the status of reason in professor Popper's scheme of things
> is, in fact, rather like that of a British civil servant, qualified
> to administer the policies of the government in power and
> even to suggest practical improvements to make them work
> better, but not to question their fundamental presupposi-
> tions or ultimate purposes.[13]

Carr's charge was to the point. And therefore if, as we have
argued, behind Bush there stood Popper, that meant a definite
restraint to human thought and action – even in the halcyon
years of post-war America.

Natural Science and Social Science

Even if Popper was right about the way *natural science* develops,
that would not imply he was right about *history*. The two are
very different. The study of nature and the study of society
therefore each require their own specific methods and
principles.

Nature has no goals and aspirations in the way that human
beings do. Humans learn – through debate, practice, mistakes –
in a way totally unlike even the brightest chimpanzee or dolphin.
Indeed, it is precisely because humanity learns so rapidly that
change in society is much more explosive than, and radically
different from, *change in nature*. Human society is also more
open to investigation and change than is nature. Society is, after
all, the creation of humanity. Nature is not.

All these differences mean that Popper erred in aligning
knowledge of nature with that of society. Popper's error hints at
a further point. While he presented his ideas on history as a
logical consequence of his ideas on natural science, the chain of
Popper's thinking in fact ran in exactly the opposite direction.
Popper's autobiography, *Unended Quest*, makes this very clear.

It was a youthful brush with Marxism, and not with the scientific textbook or laboratory, which changed Popper's life. His

encounter with the class struggles of post-1918 Austria predisposed him early on in life to deplore radical change as both utopian and dangerous. Later, Popper came to regard even the liberalism of John Stuart Mill as too extreme. Popper's trajectory is a tale of disillusionment with nineteenth-century theories of progress, and fear of their left-wing, twentieth-century successors. It is these feelings which informed his attitude to science, and not the other way round.

Thomas Kuhn and the Post-Kuhnians

Popper's intellectual divorce from progress highlighted how thinking about science had evolved a long way from that of the Enlightenment to that of Vannevar Bush's frontier without end. Once John F. Kennedy came to power and declared his country's need to open up what he termed a 'New Frontier', it was evident that further doubts had set in during little more than a decade. As America and the world plunged quickly into the climacteric of the 1960s, the writings of Thomas Kuhn presented a second fascinating example of how insecurities about social progress worked their way into thinking about natural science.

Kuhn's background was in theoretical physics. He began to study the history of scientific change at the Cornell Institute in Boston. When he published his inquiries as *The Structure of Scientific Revolutions* (1962), he created something of a revolution himself. He proposed an entirely novel way of studying scientific change.

Kuhn's deceptively simple and thus beguiling hypothesis was that the normal study of nature had been conducted within a body of scientific theory which had a coherence to it. He called this corpus of theories a 'paradigm'. Moreover, he felt that when a paradigm was unable to explain significant observations, and there seemed no way of resolving this, a crisis set in from which a new paradigm would hopefully emerge.

So far, so uncontroversial. It was Kuhn's next move that turned his *Structure* into a long-term bestseller every bit as massive in its influence as Popper's *Logic*. Kuhn argued that the influence of the paradigm was so pervasive that communication between different paradigms was impossible. It was not possible to compare notes across paradigms by reference to the way nature 'really was', because observations were paradigm-specific:

> Like the choice between competing political institutions, that between competing paradigms proves to be a choice between incompatible modes of community life. Because it has that character, the choice is not and cannot be deter-mined merely by the evaluative procedures characteristic of normal science, for these depend in part upon a particular paradigm, and that paradigm is at issue. When paradigms enter, as they must, into a debate about paradigm choice, their role is necessarily circular. Each group uses its own paradigm to argue in that paradigm's defence.[14]

Two related conclusions then flow quite logically from Kuhn's hypothesis.

First, there can be no absolute basis to knowledge. Any single truth is only truthful *relative* to the paradigm from within which it is proclaimed. Second, there is no way of saying that some paradigms are better than others. Each scientific theory need not necessarily be an advance on the last. As a sensible concept, then, progress in scientific knowledge is a non-starter.

Kuhn himself was reluctant to draw these conclusions. He was aware that to do so meant to fly in the face of powerful arguments – those of common sense. After all, Einstein's theory of relativity excellently embraces and explains observations made by and from within the previous Newtonian paradigm. Equally, Einstein's theory has many times been shown to be superior to Newton's in explaining the 'real world'. As Kuhn himself observed: 'later scientific theories are better than earlier ones for solving puzzles in the often quite different environ-ments to which they are applied.'[15]

To be fair to Kuhn, he was both aware of the problems

thrown up by his hypothesis, and also reluctant to abandon a notion of progress in science. 'We must explain', he said, 'why science – our surest example of sound knowledge – progresses as it does, and we must first find out how, in fact, it does progress.'[16] But while Kuhn shied away from the relativistic implications of his doctrine of paradigms, his successors rushed to them without shame.

The openly irrationalist writings of Paul Feyerabend, a historian of science, are a case in point. Feyerabend took ideas from both Popper and Kuhn. As far as the philosophy of science was concerned, his own comment was: 'anything goes'. Beginning as an anarchist, Feyerabend became a liberal Green who was deeply fearful of progress.

In his widely-read *Against Method* (1975), Feyerabend concluded that progress was the lodestar of a race bent on destroying the planet. He claimed that progress was not only misguided, but also impossible. For Feyerabend, true knowledge of nature was not available to humanity, because all of us are trapped inside particular frames of reference.

Coming from an earlier, slightly more optimistic generation, Popper accepted that, insofar as old theories failed and new ones passed his test of falsifiability, progress of some sort did occur in the natural sciences. In practice Popper *did* believe that there was a common world which, across the generations, scientists sought to explain. By contrast, the post-Kuhnians rejected the whole concept that the transition from one set of scientific ideas to another could be characterised as progress. They went further: they argued that the scientific method and objective knowledge played little or no role in such transitions – and certainly no role when compared with that played by politics, culture, or plain old fashion.

Made from the vantage-point of an ex-editor of the *Times Higher Education Supplement*, Peter Scott's observation on Kuhn is acute:

Kuhn's ideas have spread out across all disciplines and seeped into the broader culture of our age... They and he have matched our times. Popper, born in 1902, became the philosopher of science's last heroic age, before its progressive beneficence had been called into serious doubt. Kuhn, a generation younger, grew to maturity in the shadow of the atomic age. Maybe he has become science's court philosopher in the late twentieth-century ambivalence.[17]

Scott does, however, exaggerate the difference between Popper and Kuhn. This is in fact quite simple. Popper dismissed any notion of progress in society, but held on to the idea that scientific theories could move forward. He could not justify this; but he was relatively bullish about the future of science. Kuhn, on the other hand, was unsure even about the notion of progress in science. Distasteful though its architect found them, the Kuhnian framework was bound to lead to postures contemptuous of science.

The Atomic Era

We have now counterposed Enlightenment attitudes to those of two post-war philosophies of science. But our counterposition also holds good if post-war views on the *relationship between science and society* are subjected to closer examination. Atomic science is of moment here. For the American government, the military use of atomic science was not just *an* application of the discipline, but *the* reason for pursuing that discipline in the first place. When President Harry Truman heard that Hiroshima had been flattened on the morning of August 6 1945, he called the event the 'greatest thing in history'.[18] Equally, in the years after the war, Vannevar Bush warded off any and every movement to stop the military application of atomic science.

Bush's message was blunt: America needed the Bomb to defend democracy against the Soviets. Marshalling recent images of America as bastion against Hitler to the fresh concerns of his day, Bush said this of the discovery of the bomb:

We can never be thankful enough that the secret was learned by peace-loving peoples, not by the fascist nations which sought with all their might to master it in order to unleash atomic war on the whole world.[19]

Bush claimed that the Nazis tried 'with all their might' to develop a Bomb. But this was a plain lie: as recent historiography has shown, the Americans knew at least as early as 1944 that the Germans were not seriously trying to develop an atomic device.[20]

Bush's optimism about peace and science, then, was couched in the mendacious rhetoric of American worldwide hegemony. His was a fundamentally belligerent political schema.

Meanwhile, the practical and theoretical development of many areas of science, and not just of atomic power, was led by military objectives. American universities such as the Massachusetts Institute of Technology (MIT) and Stanford had a very close relationship with US defence contractors. As the historian Daniel Kevles relates, during this period, all roads seemed to lead to the Pentagon.[21]

In both academic establishments and a myriad of companies, research related to the military was at the cutting edge of new developments in science. The drive for military advantage in space allowed NASA to lead much of American science and to receive lavish government funding. In scores of thinktanks and at hundreds of symposia, generals and scientists rubbed shoulders as never before. Finally, through 'spin-off', a number of civilian technologies sprung, first, from developments in defence-related research.

On the whole, scientists reacted to these developments either by accepting the militaristic pattern of developments, or by indicting science in general and atomic science in particular as inherently given to abuse. Robert Oppenheimer, director of the Manhattan Project at Los Alamos, inclined to the latter view. When he saw the test explosion in the New Mexico desert, he was filled with fear. In a resonant phrase, he recalled a line from the *Bhagavad Gita*, the sacred text of the Hindus: 'I am become death, the destroyer of worlds.'[22]

By 1958, when the historian Robert Jungk published *Brighter Than a Thousand Suns*, his famous account of the Manhattan Project, Oppenheimer's moment of doubt had become, with Jungk, an explicit polemic against the Baconian idea of human mastery over nature. Jungk took it as inevitable that destruction would issue from science. He observed that Bacon's aphorism 'Knowledge is Power' had come, 13 years on from Hiroshima and Nagasaki, to be translated in the popular mind into the idea that 'Knowledge is *Unfortunately* Power'.

Citing the great American post-war physicist Richard Feynman, Jungk noticed that many scientists of his day had come to fear their God-like character. Citing Feynman once more, Jungk argued for intellectual humility in the face of the unanswerable secrets of the universe – secrets which, moreover, ought to remain unanswered. Finally, he registered the passing of an age. Hopes in progress, said Jungk, had culminated, because progress had been almost unanimously identified with science and technology, in the development of 'absolute weapons'. For Jungk as for a generation of scientists and thinkers after the Manhattan Project, that was a warning. The Bomb stood as an awful recommendation to abandon all the conquests of the Enlightenment. Because progress could no longer be guaranteed, to strive for it was dangerous, risky. The Bomb was too hot to handle; it was time to give up the struggle to dominate nature.[23]

From Oppenheimer and Jungk, we now conclude our treatment of the atomic era with Hubert Reeves and his book *The Hour of Our Delight: Cosmic Evolution, Order and Complexity* (1986).[24] When originally published in France, *The Hour of Our Delight* sold more than 200 000 copies and won the prestigious Prix Blaise Pascal. Reeves, director of research at the Centre Nationale de la Recherche Scientifique in Paris, had worked with many of the twentieth century's leading physicists, including Feynman and Hans Bethe. In a Prologue to the English edition of his book, Reeves came straight to the point about his credo:

To problems of cosmic scope I offer no grand solutions. Those presented in the past have generally failed miserably. What is needed, I believe, is a change in the state of mind of human beings. This can only take place gradually, in the context of everyday life.[25]

For Reeves, nearly 50 years after the Manhattan Project, the atomic experience is so searing that it is time not to conquer nature, but to bow down before it. Accordingly, in both science and society, Reeves finds the 'grand solutions' embarked upon by his countrymen 200 years before simply embarrassing.

Reeves holds that scientists are arrogant. They have an 'irreverence' for the workings of nature which is essentially immoral. With Reeves we have not the Frankenstein myth in relation to the life sciences, but earlier Greek myths about man not getting above his station:

Science and technology developed in the Western world precisely when the mystical relationship to nature was first questioned. This is certainly not accidental. Once again we encounter the image of Prometheus ripping the fire out of the heavens, the 'sin', according to Oppenheimer, that the physicists came to know at Los Alamos.[26]

By being intrinsically presumptuous, science is intrinsically sinful – or so Reeves would have us believe.

In the conclusion of *The Hour of Our Delight*, Reeves discusses the achievements of humanity over the past few hundred years. It turns out that only artistic successes command his affection. The discovery of DNA in his lifetime, the Copernican Revolution before it: these, and other scientific leaps forward, fail to gain a single plaudit. Reeves's omissions show that he feels that science has achieved nothing.

Looking back at the atomic era, a number of themes emerge. First, Bush and those of his political persuasion were far too inclined to blur over the use of science in the cause of progress with its abuse in the service of Second World War and Cold War objectives. On the other hand, liberal critics of the atomic era often mistook the subordination of science to American military objectives as evidence that science, scientists and humanity as a whole could no longer be trusted. Both sides showed that they had lost their grip on the relationship between science and its application by society.

The Bush camp appeared more optimistic than those dissidents broadly grouped around the *Bulletin of the Atomic Scientists*, the journal which for many years acted as the conscience of the international physics community. But as we have seen, the endless frontier proclaimed by Bush receded only a short way. In essence, Bush's vision of atomic science was of victory over Moscow.

Much of the science of the post-war period had military origins. But that fact cannot deflect us, today, from regretting the conclusions drawn by critics of the atomic era. By associating the humanist perspective of the Enlightenment with the Bushian vision, Jungk, Reeves and others perform but a sleight of hand: they equate the incomparable. In so doing, they diminish the stature of science, and narrow the terrain for future human and social progress.

Bush's talk of an endless frontier should not blind us to a further point: his vision was not only severely restricted, but also proved easily threatened by that corrosive sense of insecurity which dogged America ever since the Soviet Union's launch of Sputnik in 1957. When we turn from the atomic era to the space age, we find an American loss of certainty evident even when science and technology enjoyed perhaps the greatest upsurge this century.

The Space Age

Writing of the 1964 World's Fair, the cultural critic Andrew Ross notes:

Although the United States in 1964 was at the height of its post-War boom, in love with the Space Age, and fully subscribed to president Kennedy's New Frontier of Science and Technology, the World Fair's generic language did not hold the decisive rhetorical sway it had enjoyed in the post-Depression years of the late thirties. The resurgence of the cult of science and invention in the post-Sputnik years did not establish the same deep roots in popular consciousness as it had done in the decade before Hiroshima.[27]

Ross is right to highlight the way in which the atomic experience influenced the broad public. Yet the use of atomic weapons only had the effect that it did on perceptions of the future because much of the public, and many scientists, were pessimistic about scientific and social progress. This led to a tendency to equate the use of science with its abuse; it also predisposed popular consciousness toward a jaded view of the broader possibilities for humanity thrown up by advances in science. One way of seeing just how disorientated people can get about science is to examine attitudes to the Space Age.

While there was enthusiasm for the exploration of space, the coming of the Space Age to America also brought to the surface serious national anxieties.

The race for the moon provides, in the first place, a good example of the military's domination over science in the post-war period. NASA was only formed after the Soviet Union beat America into space with Sputnik 1. America was motivated to go to the moon to shore up its faltering position on earth. As the historian Walter McDougal reminds us in his *The Heavens and the Earth: A Political History of the Space Age* (1985), after Sputnik, the French physicist Frédéric Joliot-Curie, an admirer of the Soviet Union, declared:

It is no accident that the Soviet Union was the first to launch a Sputnik. In this fact lies the law of development of the new society. This outstripping of Western science will each year become more frequent.[28]

As McDougal suggests, this was a challenge the American élite could not ignore. In 1960, Lyndon Johnson declared that it was 'not an overestimate to say that space has become for many people the primary symbol of world leadership.'[29] Behind America's drive to get back into the space race lay the haunting suspicion of failure. Indeed, the very idea of a 'race' spoke of a loss of ground; even, perhaps, of a loss of American direction.

In retrospect, the launch-pad for American efforts in space lay on *terra firma*. The recession of 1957–8, the first downturn since the Korean arms build-up of 1950, showed that the American economy was not invincible. During those years, commentators from Vance Packard through to the respectable press worried whether the terrible 1930s might return. In the 1958 Congressional elections, the Democrats routed the Republicans. *The Wall Street Journal* said the results reflected a feeling of 'impending disaster'. The loss of confidence in Republican President Eisenhower, said one commentator, reflected 'a sense that as a nation we are beset by problems which are slipping beyond our control.'[30]

As the 1960s opened, Soviet successes in space and elsewhere became the prism through which American neuroses were expressed. The crisis of credibility of the period brought Kennedy into the White House and led him to undertake the Apollo programme. Kennedy was brought round to the programme after Yuri Gagarin went into orbit in April 1961 and, over Cuba, the American Government was humiliated in the Bay of Pigs fiasco. Was there any space programme, asked Kennedy of vice-President Johnson, which promised 'dramatic results' and which at the same time America could win?[31]

America's rulers knew that something – anything – had to be done, fast. But if their worries were national, the space race also generated popular feelings of concern. Officially a triumph, Apollo did enthuse the American public. Quite soon, however, another sentiment came to bear: for space had shown that human ambitions should perhaps be scaled down.

To the burgeoning ecology movements of the years 1965–75, the image of Earth from space betokened the smallness of humanity, the fragility of the Earth, and the awesome character of the cosmos. The vastness of space, boundless territory to explore, endless secrets to unravel and – one day, perhaps – limitless resources to mine: this became the vision of a limited few. For many more, space confirmed that humanity suffered from a severe case of delusions of grandeur. Walter McDougal draws this lesson. 'No matter how intimidating (to ourselves) man's power may grow', that power 'will always be picayune beside that of nature.'[32]

Shifting Perceptions of Science

Over the centuries, the shift in perceptions of science has been dramatic. Condorcet, let us remind ourselves, did not know the car, and yet dreamt of human control of nature. By contrast, contemporary historians have seen a man on the moon, but want to kneel before nature and condemn human beings as flawed from the start.

A number of commentators on post-war affairs have tracked the ways in which technological advance has coincided with negative perceptions of science and progress. Taking an overview of the period, Peter Scott acutely observes:

> Our claims were modest compared with those made by the ideological giants of the nineteenth and early twentieth century; at the same time, our scientific achievements were immensely more impressive than theirs. The more we shrank from offering grand explanations, the more effective became our technology.[33]

All that needs to be added to this is that the balance between optimism and insecurity at the time, and the relationship of trends then to contemporary trends, needs further analysis.

Whatever the motivations of post-war Western Governments, and despite the destructive results of those motivations, there was a public support for science and technology in the post-war period that is largely absent today. There was hope, for instance, that civil nuclear power would bring cheap electricity for all. There was also outright euphoria about going to the moon. Though JFK's promise of the new frontier soon proved chimerical, prosperity in the West buoyed up popular appreciation of science. This feeling, combined with a greater sense that problems could be solved, gave the period a more optimistic tone than that which followed.

Indeed it was only when, in the 1970s, post-war economic boom gave way to stagnation and mass unemployment that large-scale scientific and social projects came to be seen as folly. From then on, however, the destructive consequences of the way in which society applied all forms of technology was widely taken to be the inevitable consequence of any ambitious large-scale project. Small, as Ernst Schumacher pronounced, was beautiful... in a way that big could never be.

In his brilliant *All That is Solid Melts into Air: The Experience of Modernity*, the radical New York cultural and literary critic Marshall Berman documents the way in which large urban projects in the USA were soon seen as devastating rather than potentially creative. 'To be modern', Berman acknowledges rather ruefully, 'turned out to be far more problematic, and more perilous, than I had been taught.'[34] Through the 1960s and 1970s, a 'morning after' sensation also set in over science and technology in the Third World. The collapse of African economies, and of some of the major construction projects they had earlier undertaken, seemed to confirm Schumacher's manifesto, and in particular his call for a lowering of horizons. The cumulative impact of developments such as these now means that, by the closing years of the twentieth century, even the qualified but nevertheless perceptible optimism of the 1950s and 1960s seems a world away.

The general perception of science and progress which is prevalent today is one of impending tragedy. As a result, the attitudes toward science expressed after the war are ridiculed as naïve, if not utopian. Adam Curtis' *Pandora's Box*, six widely-acclaimed and polished 'fables from the age of science' shown on British television in 1992, summed up the mood of the early

1990s very well. The series lampooned sinister American technocrats of the 1960s. But it did not explain their bellicose ravings by any reference to the USA's need to show the world that it could, after all, prevail against Vietnam. Rather, the BBC programmes identified the Pentagon's lunatic fringe as the necessary outcome of a scientific conception of the world. In the same spirit, they portrayed the DDT insecticide programmes of the 1950s as the outcome not of scientific ignorance, but of a surfeit of science: 'we thought', a US farmer dutifully admitted, 'we were building the American Dream'.

The consensus against progress today is in part a reaction to the failed project of post-war modernisation, and to that project's frequently devastating use of science and technology. At the same time, contemporary critics of science and progress merely confirm that post-war endorsements of science and progress were always ambiguous – always capable of falling victim to that wider sense of social malaise which has now engulfed Western society.

Three Theories Mobilised against Progress

One thing which distinguishes our times from the post-war decades is that science itself has now been conscripted into untoward campaigns. Quantum mechanics, chaos theory and theories of complexity are now widely taken to imply that human ambitions must necessarily be diminished by nature and by the complexity of interactions within it. By contrast to the Enlightenment vision of humanity as superior to nature, the new science is taken to imply that human beings must concede to nature, since they are dependent upon it, and cannot ever fathom – still less take genuine advantage from – its complexities. Quantum mechanics and chaos theory, we are told, 'launch an assault on the mechanistic reductionism that has robbed the Western world of spiritual meaning for over three hundred years.'[35]

In the realm of wealth creation it is a similar story. There, the rules of chaotic and complex systems have been cited as the reason why governments can do nothing except accommodate to market forces. From London, the Institute of Economic Affairs, a right-wing thinktank, in this vein celebrates the anarchy of markets. Being chaotic, markets

> cannot be driven to realise anyone's prior intention. Instead, such systems evolve through a process of self-organisation from which their futures emerge. Members of such a system contribute to its unfolding future, but none can be in control of it.[36]

Here humanity's subordination to supply and demand is accepted as a scientific fact of life.

At New Mexico's Santa Fe Institute, John Casti, executive editor of the journal *Complexity*, draws similarly broad and bleak conclusions:

> To those raised in the suffocating embrace of a can-do, the impossible-just-takes-a-little-longer culture, with its emphasis on an almost mystical belief in the powers of the human spirit and mind to overcome virtually any obstacle, a tour of twentieth-century science must be a quite discouraging and depressing experience. If there's any message that modern science can be confident in trumpeting to the world, it's the sobering thought that there are limits – even to the human spirit... Taken as a whole, modern science has redrawn the map of human knowledge so that it now shows potholes and detours not only along every side street and back alley, but on all the major highways and byways as well.[37]

For Casti, science is testimony to human weakness. Ecologists too have rushed to welcome these ideas. At New York's Environmental Defence Fund, Bruce Rich, a director, argues that the 'profound heterogeneity' of ecosystems, economies and societies, conforms to the rules outlined by complexity theories. That, in his view, makes attempts to know, predict, plan, and manage 'global environmental crises' about as redundant as 'a predictable Newtonian-Cartesian model.'[38] Rich demands a total re-think of social practice:

> We are in the midst of a momentous historical transition in which the world we have created through three centuries of modernity is now so changed that it is forcing us to abandon its epistemological, cultural, and political assumptions. In such a period of change and uncertainty, values such as preserving future options and diversity, and conserving the robustness of ecosystems, local communities, and institutions assume extraordinary importance.[39]

Like many other environmentalists, Rich believes that science has now proved, incontrovertibly, that it is time to reduce ambitions, stop thinking global and start – and indeed end – by acting local.

Altogether, quantum mechanics, chaos and complexity form three areas of modern science which are widely claimed to invalidate the goal of human mastery of nature. The next three chapters of this book examine the validity of that claim.

Chapter Two:
Belittling Humanity

'There are limits – even to the human spirit', says John Casti. We have to abandon Modernity's 'epistemological, cultural, and political assumptions', according to Bruce Rich.

Rich bases his point on complexity theory. Casti relies on quantum, chaos and complexity theories. Our quotations are but a sample from an extensive literature which restricts endeavour and belittles humanity.

This chapter serves as an introduction to the following two chapters, which examine quantum, chaos and complexity theories. We here adopt the rather unusual procedure of spelling out the conclusions drawn from the new science, and giving an outline of the scientific issues at stake, *before* discussing the detailed science. We do this to show why the reader should persevere with two rather more technical chapters. More importantly, we want to focus attention on the issue of *interpretation*.

We take issue with the conclusions Casti, Rich and many others draw. These conclusions, which concern human abilities, are drawn from a growing body of scientific knowledge generated by many leading scientists. But are the interpretations valid? And if not, where are the illegitimate steps taken? This chapter sets a framework for pursuing these issues in Chapters Three and Four.

Causality, Purpose and What it is to be Human

Underlying the Enlightenment conception of progress is the idea that humanity can mould nature. This view rests on three key assumptions: that humanity can manipulate cause and effect in nature to its benefit; that purpose or intention are imposed on nature by humans and are not a feature of nature itself; and that society is both separate from nature and a means by which people can impose their will upon nature.

In combination, quantum, chaos and complexity theories are now widely taken to undermine all three Enlightenment assumptions.

The three assumptions owed much to ideas developed by Bacon. For him, natural law was governed by underlying mechanical causal interactions, and nothing more. This view of the world was also taken up by Descartes, who is usually seen as the pioneer of the mechanistic approach, since he explicitly put forward a purely mechanical view of all non-human processes – including animal and plant life.

Bacon linked his outlook on natural law to his view that humanity should be the master of nature. He argued that a knowledge of the mechanisms, of the series of causes and effects, that were to be found in nature would enable men to turn it to their own ends. He wrote:

Human knowledge and human power meet in one; for when the cause is not known the effect cannot be produced. Nature to be commanded must be obeyed; and that which in contemplation is as the cause is in operation as the rule.[1]

Bacon did little theoretical or practical science himself. Descartes did postulate some mechanical models of various things, which, with the benefit of hindsight, were clearly no more than vivid imaginings, and rather simplistic ones at that. A more sophisticated view of natural law based on Bacon's assumptions was put forward this century by Albert Einstein. In a modified form, it lies behind the approach to the study of natural law taken by Steven Weinberg and many other theorists involved in the study of particle physics today.[2]

For Bacon, the imposition of human will upon nature was

made possible not only through a knowledge of cause and effect, but also by the fact that there was no plan to nature's workings. Bacon developed this view in opposition to *animism*, which held that both living and non-living matter obeyed similar laws, and that there was a wider, lifelike, organising principle present in both. Animism was often presented in the form of teleology, which posited a goal, or final cause, for the universe.

It was Aristotle [384–322 BC] who had pioneered the animist approach. In his *Physics*, Aristotle had identified four complete and mutually exclusive categories of causality: material, efficient, formal, and final causation. Each category answered a different 'why' question. Aristotle had asked, for example: 'Why is my house the way it is?'. His answer had been that it is the way it is because of the materials from which it is constructed (material cause), because of the labour that the workmen put into it (efficient cause), because of the architectural plans describing the structure of the house (formal cause), and because of his desire to have a comfortable shelter from the elements (final cause).[3]

This is a sensible way to look at the construction of houses. But is it a sensible way to view nature? While Aristotle thought it was, Bacon disagreed. Bacon rejected any notions of final causation in nature, and the purpose which they implied. Purpose, suggested Bacon, is a uniquely human thing. We design a house to live in because we are conscious. Bacon also pointed out that positing a final cause to a natural process did not actually explain anything that was not already known. In his *Age of Discovery*, he wrote: 'Inquiry into final causes is sterile, and, like a virgin consecrated to God, produces nothing'.[4]

That there is no underlying purpose or direction to nature means that humanity is free to exert its preferences on nature. But how can humanity have the tools, the means to carry this out? Our view is that *social organisation* gives humanity the power to reconfigure nature.

It was implicit, and sometimes explicit in Enlightenment thinking that social organisation was distinct from nature. For the Enlightenment, humanity's powers originated precisely in this distinctiveness. Of course, humanity is a part of nature. However, social organisation has raised humanity above the level of animals. In so doing it has conferred upon men and women the ability to tame nature. If, at times, Enlightenment thinkers understated the uniqueness of humanity, their more crucial weakness was that they could provide no explanation of this uniqueness that did not rely on God's beneficence.

The Challenge Posed by Quantum Mechanics

The idea that cause and effect in nature can be known and put to work is, in the opinion of many, undermined by quantum mechanics, the origins of which go back to the year 1900. Quantum mechanics is one of the most successful theories that science has produced to date. It has helped to unravel the structure of the atom, explain the nature of the chemical bond, and predict the existence of anti-matter. At a more practical level, quantum mechanics has been central to the development of electronics, from the transistor to the microprocessor.

For many of its originators, however, the new theory also challenged classical Newtonian conceptions of nature. The Danish physicist Niels Bohr, a major architect of quantum theory, asserted that anyone who was not shocked by it had not understood it.[5] Many among an older generation of physicists certainly were shocked. Not only was the new mechanics difficult for them to comprehend, it also seemed to fault their underlying assumptions about the character of science. Nobel Laureate Hendrik Lorentz, for example, wrote in the late 1920s:

> I have lost the conviction that my work has led to objective truth, and I don't know why I have lived. I only wish I had died five years ago, when everything seemed clear to me.[6]

In the twentieth century, the consensus has grown that quantum mechanics undermines materialism – the view that everything in the universe consists of matter, in different forms – and realism – the view that nature has an objectivity independent of human consciousness.

Quantum theory, it is held, subverts everything that has gone before. Arthur Koestler's comments here are typical. In 1972, he wrote:

> It is time for us to draw the lesson from twentieth-century post-mechanistic science, and get out of the straight-jacket which nineteenth-century materialism imposed on our philosophical outlook. Paradoxically, had that outlook kept abreast with modern science itself, instead of lagging a century behind it, we would have been liberated from that straight-jacket long ago.[7]

The question which Koestler begs, nevertheless, is how his 'lesson' from post-mechanistic science can be drawn from quantum mechanics. Equally, we must ask if Bohr's own startling conclusions from the new theory – that discontinuity replaces continuity in nature, and that indeterminism and acausality reign supreme – are fairly drawn.

Continuity and Causality

The notion of continuity underlying the classical description of physical phenomena can be illustrated using temperature. The temperature of an object varies in a smooth and continuous way. There is no sudden jump from one temperature to another without passing through all the intermediate values. For Bohr, this Newtonian assumption of properties varying continuously is not applicable in the 'micro' world of the atomic realm. Here quantum theory replaces continuity with discontinuity and discreteness. The very word 'quantum' means 'a discrete amount' or 'packet'.

In addition to continuity, the world-view of classical physics had, as we have seen, a strong commitment to universal causality. To recapitulate, it is worth quoting the most famous expression of this outlook – that given by the Marquis de Laplace [1749–1827] in his book *A Philosophical Essay on Probabilities* (1816):

> We ought to regard the present state of the universe as the effect of its anterior state and as cause of the one that is to follow. Given for one instant an intelligence which could comprehend all the forces by which nature is animated and the respective situation of all beings who compose it – an intelligence sufficiently vast to submit these data to analysis – it would embrace in the same formula the movements of the greatest bodies of the universe and those of the lightest atom; for it, nothing would be uncertain and the future, as the past, would be present to its eyes. The human mind offers, in the perfection which it has been able to give to astronomy, a feeble idea of this intelligence. Its discoveries in mechanics and geometry, added to that of universal gravity, have enabled it to comprehend in the same analytical expressions the past and the future states of the world.[8]

In the classical physics of Laplace and others, a particle has a definite momentum and position at any given time. The forces that act on the particle determine the way its momentum and position vary with time. If we consider a gas, which consists of many such particles, we have to introduce probability and settle for a statistical description. This retreat into a statistical analysis follows, said Laplace, from the lack of 'an intelligence which could comprehend all the forces by which nature is animated' and the lack of 'an intelligence sufficiently vast to submit these data to analysis'. However, for the classical physicist the physical world remained determinate; probability was not an intrinsic feature of the world. Rather, it was a consequence of human ignorance.

Quantum mechanics, however, introduced indeterminacy, probability and statistical descriptions as essential features of physical phenomena, rather than as artificial constructs made necessary by incomplete information. The simplest example of this is the quantum-mechanical interpretation of radioactive decay. That interpretation says that the decay of an individual nucleus is a phenomenon without a cause. The theoretical physicist Paul Davies uses this to make the wider case for acausality at the atomic level:

> Quantum fluctuations are not the result of human limita-tions or hidden degrees of freedom; they are inherent in the workings of nature on an atomic scale. For example, the exact moment of decay of a particular radioactive nucleus is intrinsically uncertain. An element of genuine unpre-dictability is thus injected into nature.[9]

The German physicist Werner Heisenberg went even further. Generalising from the atomic domain to the 'macro' world, he laid out the principle that quantum mechanics 'definitely estab-lishes the non-validity of the law of causality.'[10]

The acausality and indeterminism which Bohr wrote into the very character of physical law is the reason why quantum theory is so often invoked today. These phenomena seem to bar humanity from that knowledge of causes and effects in nature which is central to any and all efforts to re-order the physical world. But Bohr's work also challenged another key tenet of Enlightenment theory and classical science: the separation of subject (humanity) and object (nature).

The effect of the interaction between an electron (the object) and the observer (the subject, which in quantum mechanics can be a piece of laboratory equipment as much as a human observer) cannot be ignored when an observation is made. According to Bohr, we cannot speak of nature having properties distinct from our observational framework. There is no independent 'there' which can be truly said to be 'out there'.

Pulling the two sides of his doctrine together, Bohr wrote that quantum mechanics implies 'a final renunciation of the classical ideal of causality and a radical revision of our attitude toward the problem of physical reality.'[11] But such a result is of course profoundly counter-intuitive. We all assume that nature has a definite character, with definite rules governing cause and effect. Moreover, it is not just lay-people that think this. In their practical work, physicists behave as if there is a physical reality 'out there' with definite properties, which can be investigated. Faced with Bohr's conclusions, most physicists muddle through by separating their practical work from the oddities of quantum mechanics. As the astronomer John Barrow points out:

> Almost every working scientist is a realist – at least during working hours. Although, if he is honest, he has probably never given the matter much thought, because he has found his investigations and researches to be almost entirely independent of his views, yet if questioned at the weekend he might not wish to defend the realist position too strongly.[12]

The pragmatic approach Barrow outlines allows the practical physicist to get by. But it is no solution to the wider issues at stake. More brutally, it is hard to envisage progress based on mastery of nature when the very objectivity of that nature is anyone's guess.

The conclusions Bohr drew are the ones that are believed to set limits to human possibilities. His *interpretation* is the most popular one, and is taken by many to be quantum mechanics. And yet the interpretation of quantum mechanics which Bohr and others drew is itself open to challenge. Einstein was one of the founders of quantum theory. Later, he wrote of it: 'the more success the quantum theory has, the sillier it looks.'[13] He objected in particular to the claim that quantum mechanics overthrew objectivity and causality in nature. For Einstein, the universe was wholly imbued with causality. Causality, too, was not just universal: it existed independently of human consciousness.

These ideas were, for Einstein, central to the rational pursuit of science. On the bicentenary of Newton's death, he wrote to the secretary of the Royal Society: 'may the spirit of Newton's method give us the power to restore unison between physical reality and the profoundest characteristic of Newton's teaching – strict causality.'[14] On another occasion Einstein wrote that if he had to give up causality as a result of quantum mechanics, he would rather be a cobbler, or even an employee in a gaming house, than be a physicist.[15]

There can be no doubt that quantum mechanics captures a profound aspect of physical reality. What deserves further scrutiny, however, is the relationship between the theory, and the *interpretation* of that theory which Bohr drew.

Chaos Theory and the Unpredictability of the 'Macro' World

Some scientists are prepared to accept Bohr's renunciation of causality and realism, but only in the atomic domain. They refuse to draw his broader conclusions about the physical world at the 'macro' level. Perhaps, they say, the disquieting results of quantum mechanics can be confined to the atomic realm.

Chaos theory now seems to destroy even this forlorn refuge. According to the dominant *interpretation* of chaos theory, there can be no return to the certainties of the pre-quantum world. As the introduction to the NEW SCIENTIST *Guide to Chaos* explains, the 'macro' world 'may have its own uncertainty principle, a result of the nature of chaotic dynamics rather than quantum probability.'[16]

Converts to chaos theory tend to exhibit a zeal about it. They see in the theory's non-linear equations a driving principle for all natural and social events – a principle which destroys predictability in every system it touches. In his widely acclaimed *Does God Play Dice?*, the mathematician Ian Stewart proposes that chaos theory is 'a whole new world of ideas and possibilities, a new kind of mathematics, a fundamental insight into nature itself, and it brings us closer to an understanding of literally everything.' But because of his desire to apply chaos theory to every phenomenon, Stewart can only conclude that there are limits to what human beings can know. That is why the final chapter of his book is very apposite: 'Farewell, Deep Thought'.

Ignorance is one thing. But most writers on the application of chaos theory delight in the impotence it allegedly confers. In an article on chaos in medicine, we are told: 'it follows that we patients should not expect too much of our doctors!'.[17] A study of chaos and capital markets by Edgar Peters concludes: 'we can no longer solve for optimal solutions, but must instead be content to examine probabilities in a world that can abruptly change when certain critical points are passed. This new view of the world offers us less control over our environment, even as it offers us a more complete picture of how the world works.'[18]

If natural and social systems are fundamentally unpredictable, as the zealots of chaos theory argue, then they are certainly uncontrollable. Chaos theory is widely taken to assist quantum mechanics in undermining the Enlightenment project. As far as the popular interpretation of the theory is concerned, however, the word 'chaos' is talismanic enough to meet all objections. The London style magazine *i-D* exemplifies this very well:

> If you want another label you could call chaos post-industrial, post-modern science. The basic assumptions of the deterministic project of modern science... have been undermined by chaos. The new science is more responsive, less arrogant. The arid, totalizing, almost totalitarian assumptions of modern science have been modified by the reticent appreciation of the complexity and unpredictability of real life.[19]

The one thing discussions of chaos theory can always be relied upon to generate is sage nods of agreement.

'Chaos' conjures up images of things falling apart, of things descending into... well, chaos. As such the term admirably sums up the times in which we are living. It is little surprise that the American management guru Peter Drucker gave a warm welcome to chaos theory as early as 1989, in his book *The New Realities*. For Drucker, chaos made government economic policy redundant. Governments should just stand back.

The promiscuous use of the term 'chaos' can confuse the real issues at stake. Mathematically, chaos is a phenomenon associated with a certain kind of mathematics termed non-linear dynamics. It needs to be distinguished from simple disorder, whether that exists in nature or in society. Disorder in nature and society may be actually chaotic in character, or it may just be the complicated interaction of many phenomena. Detailed investigation is needed to establish the nature of disorder in each particular circumstance.

Unlike quantum mechanics, the basic equations involved in non-linear systems are very classical. Indeed, the theory is totally deterministic and causal. There are none of the weird acausal happenings we find in the quantum domain. All the more amazing, then, that chaos theory tells us that prediction is dead. It is the claim that chaos theory kills prediction in the macro domain, despite its deterministic origins, which leads many of its exponents to claim that the new theory has caused a revolution in science.

How does chaos theory kill prediction? By showing that the evolution of a system governed by chaos is inordinately sensitive to the starting-point of the system: so sensitive, in fact, that no matter how small the difference between two different starting-points, the outcomes will, in the end, diverge completely.

This super-sensitivity has epistemological consequences. In practice, if we study a real dynamical system, we inevitably make some error in measuring the starting-point. But according to chaos theory, that modest error kills prediction. Indeed, many chaos theorists also deny that the problem is simply the practical difficulty of measurement. No matter how small the error made, they say, it cannot be contained. In this sense the problem is as much theoretical as it is practical.

Chance, Chaos and the Magnification of Differences

Interestingly enough, the birth of chaos theory can be traced back to the same period that saw the emergence of the quantum concept – the turn of the century. The origins of quantum theory lie in the Germanic world. Chaos, by contrast, is very much a French affair.[20]

In his history of the subject, the acclaimed French mathematician David Ruelle marks out a line of French mathematicians which, by the 1920s, had developed the essentials of the mathematics of chaos theory. At the end of the nineteenth century, the mathematician Jacques Hadamard discovered the phenomenon of super-sensitivity to initial conditions. Building on his work, the historian and philosopher of science Pierre Duhem spoke in 1906 of an 'example of a mathematical deduction forever unusable'.[21]

It was, however, Henri Poincaré who really appreciated the shocking nature of the subject. Poincaré, like his mathematical quarry, Laplace, was an enthusiastic player of roulette and other games of chance at the casino tables of Paris. It was through a comparison with 'chance' happenings that Poincaré realised what others had failed to see. In his famous *Science and Method* (1905), he wrote:

> Chance is only the measure of our ignorance. Fortuitous phenomena are, by definition, those whose laws we are ignorant of. But is this definition very satisfactory?[22]

Poincaré thought that the definition was not. Instead, he postulated that there was some regularity underlying unpredictability:

> It might be that small differences in the initial conditions produce very great ones in the final phenomena. A small error in the former will produce an enormous error in the latter. Prediction becomes impossible, and we have the fortuitous phenomenon.[23]

Ruelle summarises Poincaré's reconsideration of chance thus: 'a very small cause, which escapes us, determines a considerable effect which we cannot ignore, and then we say that this effect is due to chance.'[24]

For more than 50 years Poincaré's ideas lay dormant. Efforts were devoted more to quantum mechanics and other areas of physics. Then, in the 1960s, while working on the mathematics of weather prediction, Edward Lorenz stumbled across the kind of phenomena which had exercised Poincaré. During the 1970s, the mainstream mathematics and physics community largely ignored the new ideas. Practitioners of chaos theory were typically physicists on the west coast of America, happy to do their own thing.[25] But today, chaos theory is very much in the mainstream. It is taught in many universities, and features prominently in scientific research. Indeed, in 1991, the related subject of *fractals* had become so popular that it appeared in no fewer than a third of the papers submitted to the leading American physics journal *Physical Review Letters*.[26]

If quantum mechanics does not put an end to the Enlightenment project, chaos certainly does, its enthusiasts proclaim. But the real point at issue is whether they are right to contend that non-linear mathematics provides a new, *universal* theory of natural and social systems. Chaos theory is maths; but how much is it physics, chemistry, a theory of society, and so on?

The origins of the theory advise in favour of scepticism. By comparison with quantum theory, they are very mathematical. Physical and social models built upon chaos theory came well after its first introduction, and even today often lack substantiation. Indeed, David Ruelle thinks that the mathematics of chaos theory may be compromised by their association with the physics of chaos. The latter, he thinks, has produced *poorer* results, rather than better, as more and more people have come into the field:

the field has been invaded by swarms of people who are attracted by success, rather than by the ideas involved. And this changes the intellectual atmosphere for the worse... the mathematical theory of differential dynamical systems has benefited from the influx of 'chaotic' ideas and, on the whole, has not suffered from the current evolution (the technical difficulties of mathematics make cheating hard).

The physics of chaos, however, in spite of frequent triumphant announcements of 'novel' breakthroughs, has had a declining output of interesting discoveries. Hopefully, when the craze is over, a sober appraisal of the difficulties of the subject will result in a new wave of high quality results.[27]

Complexity Theory: A New Approach to Order in Nature

The study of complexity, or order, as the subject is often known, is a growing field.[28] Pierre-Gilles de Gennes was awarded the 1991 Nobel Prize in physics for work on practical applications of complexity theory which simultaneously developed some of the theoretical principles.[29] The search is now on for a theoretical synthesis of research in different fields of the natural sciences and, perhaps, the social sciences too. John Barrow anticipates that

> the very diversity of such organised complexity hints that it might be possible to abstract the very notion of complexity from the specific manifestations of it that we witness, and search for some general principles which govern its emergence and development.[30]

There are a number of competing perspectives on what might constitute a general theory of complexity.

Non-linear dynamics contains a profound theory of order, and provides one way of looking at the problem of complexity. There is a growing belief that universal principles governing the emergence of order in nature and society will be found through the use of non-linear mathematical models. The search for universal principles has always guided one of the pioneers in the modern study of complexity theory, Ilya Prigogine of the Free University of Brussels.

Prigogine's explicit aim is to create a *general* theory. He wants to explain all forms of order. In awarding him the 1977 Nobel Prize for chemistry, the Nobel Committee observed that he had created theories to 'bridge the gap' between 'the biological and the social scientific fields of inquiry'.[31]

The claim that a universal theory can explain both natural and social systems is a counter to the Enlightenment conception of progress, which, as we have seen, implicitly assumed that social organisation was different from the organisation of nature. But Prigogine's theory goes further, because it implies that order in society is no more the result of human action than is order in nature. Both, for Prigogine, are examples of ordering processes inherent in the dynamics of systems.

Prigogine contrasts his views not only with those of the Enlightenment, but also with Existentialism. In emphasising human uniqueness and freedom of choice, Existentialism, for Prigogine, believes that 'man's place in the universe is what he makes of it'.[32] For Prigogine, by contrast, human freedom is restricted by two significant facts. First, social organisation is highly dependent on natural systems. Second, it is a product of ordering principles which lie beyond the scope of human intervention.

Similar themes are taken up by Murray Gell-Mann, a Nobel Laureate in physics. Gell-Mann spends part of his time studying complexity at the Santa Fe Institute. His view is that 'the distinction between nature and human culture is not a sharp one; we human beings need to remember that we are a part of nature'.[33] The approach is premised on a theory of order which brackets together all human and some natural systems on one side, and non-living systems such as galaxies on the other. It leads Gell-Mann to grant a new urgency to environmental conservation. For if humanity is but a part of one interconnected complex system, disturbance to ecological systems could spell doom for *homo sapiens*.

At the end of Chapter Four, we outline why we think Prigogine's and Gell-Mann's approach to understanding the relationship between nature and society is fundamentally wrong. For the moment, however, we are concerned to examine complexity theory as a theory of nature, and to see how, if at all, it shatters the Enlightenment vision. The views of French Nobel

Prize winner Jacques Monod – views which Prigogine and others dispute – are worth discussing in this context.

Monod presented a Darwinian conception of nature in his book *Chance and Necessity* (1970). There he argued that the origin of life was a chance occurrence: that the developments which led to the first primitive self-replicators, the forerunners of DNA, were accidental. From then on, Monod continued, evolution was a mixture of the random mutation of genetic information, and the necessary development of mutations into living organisms – organisms which may or may not have an evolutionary advantage, depending on the mutation and the environment in which they live.

For Prigogine, on the other hand, chance at most realises a process which is already given. Likewise, Hubert Reeves sees a role for chance only in the 'actuation' of matter's 'immense potentialities'.[34]

By dethroning chance, some advocates of complexity theory move rapidly to the view that *humanity's existence is pre-ordained*. They refuse to believe that nature is simply the blind interaction of cause and effect. In this vein Paul Davies concludes his *The Mind of God: Science and the Search for Ultimate Meaning* with these words:

> The physical species homo may count for nothing, but the existence of mind in some organism on some planet in the universe is surely a fact of fundamental significance. Through conscious beings the universe has generated self-awareness. This can be no trivial detail, no minor byproduct of mindless, purposeless forces. We are truly meant to be here.[35]

Many of the ambassadors for complexity theory take a similar view, without being so explicitly religious. Prigogine opines that 'our connection to time and irreversibility gives more meaning to human life by showing that we are embedded in something we see also in us'.[36] At the Santa Fe Institute, Stuart Kauffman writes: 'life is the natural expression of complex matter. It's a very deep property of chemistry and catalysis and being far from equilibrium. And that means that we're at home in the universe, we're to be expected.'[37] As John Barrow quite rightly notes, a teleological, Aristotelian perspective has re-emerged in the study of complexity in the abstract – that is, as a general phenomenon not necessarily tied to a particular complicated physical situation.[38]

Complexity theory is therefore widely seen to challenge Enlightenment notions because it is taken not only to undermine the uniqueness of human society, but also to imply that nature is goal-directed, or teleological. That, as Brian Arthur of the Santa Fe Institute has sympathised, is 'not at all new to Eastern philosophy'. Humanity's best bet, for Arthur, is just to 'quietly observe the flow, realising that you're part of it, realising that the flow is ever-changing and always leading to new complexities'. The best humans can do in this circumstance is, every so often, to 'stick an oar into the river and punt' from one eddy to another.[39]

The first problem which needs to be raised about the validity of this new approach to order concerns its scientific relationship to Darwinism. All the proponents of a 'principle of complexity', and especially those who follow Prigogine, see nature as in some sense a 'whole', working in a cooperative way and governed by an underlying process of complexification. All of them play down Darwin's emphasis on the role of chance in evolution, even if, given Darwin's stature, few openly say that Darwin was wrong. Nevertheless, their views can only contradict Darwinism.

There can be no 'holistic' version of Darwin. His theory of *chance* mutation rules this out: no species can evolve for the good of another in any way, since its modification down the generations is shaped by chance happenings and what is good for it, not some other species, and certainly not the whole biosphere. If one organism were to evolve for the good of another, 'it would', said Darwin, 'annihilate my theory'.[40]

A second problem with complexity theory revolves around humanism. Enthusiasts for complexity theory often claim that it is a humanistic theory because it sees life, and ultimately human life, as an inevitable development. Thus for Paul Davies, complexity theory is confirmation of cosmic design.[41] Darwinism, by contrast, is presented as a bleak theory; 'heroic pessimism' was the description of Erich Jantsch, a follower of Prigogine.

That humanity might counterpose itself to nature, complexity's partisans reject. They favour an older, and meeker, vision of humanity's relationship to nature. But in robbing humanity of freedom to manipulate nature, they deny it choice. Is it not their theory, therefore, rather than Darwinism, which presents a bleak picture of the human condition? After all, what kind of future is it to 'go with the flow' – occasionally sticking an oar in, as Brian Arthur would have us do?

Contrasting views of humanism would count for very little, however, if there was overwhelming evidence for the universal application of complexity theory. Here Prigogine holds that his 'process'-oriented view of natural law has reached the status of solid science. As he puts it:

In classical physics you really had no choice. You either had to accept an alienating science, which men like Monod... expressed very clearly; or you had to go to an anti-scientific, metaphysical view, such as that of Whitehead or Bergson. Today things come closer together.[42]

But what does 'come closer together' mean? Some writers have argued that Prigogine's work proves the truth of Whitehead's metaphysics and of Bergson's mystical vitalism.[43] In the influential *Phenomenon of Man* (1955), the Catholic theologian Pierre Teilhard de Chardin used terms such as 'principle of complexification' before they were taken up by the scientists. And de Chardin also said that his thought was scientific, not religious.

Is Prigogine's work science or religion? Is the claim that complexity is a universal theory of nature and society motivated by scientific insight? Or is the motor behind the claim a liking for the conclusions that are taken to follow if the claim holds good?

In their varied ways, the dominant interpretations of quantum mechanics, chaos and complexity set limits to human ambition. All belittle humanity. The next two chapters critically examine the science, and the conclusions drawn, in an attempt to separate scientific fact from interpretation.

Chapter Three:
Quantum Mechanics and the 'Risky Game'

Quantum mechanics is simultaneously the most powerful and the most intriguing physical theory of the twentieth century. Without it, most of twentieth-century science would not exist. Yet quantum mechanics also features prominently in an endless list of books which mix science and mysticism in the same blending machine. How can this be? Murray Gell-Mann expresses the contemporary bafflement by describing quantum mechanics as 'that mysterious, confusing discipline which none of us really understands but which we know how to use'.[1]

The vast majority of physicists take the advice of Gell-Mann's fellow Nobel Laureate, Richard Feynman, who recommended that they stop trying to fathom out the basic meaning of quantum mechanics. Feynman counselled his peers just to settle for using quantum mechanics as a powerful computational tool:

> I think I can safely say that no-one understands quantum mechanics... Do not keep asking yourself, if you can possibly avoid it, 'but how can it be like that?', because you will get 'down the drain', into a blind alley from which nobody has yet escaped. Nobody knows how it can be like that.[2]

Why should two of the most brilliant physicists of the twentieth century not understand one of the great scientific theories of modern science? What kind of theory can be so difficult that nobody understands it?

Quantum mechanics is not terribly difficult to understand

mathematically. Any graduate student of theoretical physics is well versed in its mathematical structure and in its power when applied. The problem that lies at the centre of quantum mechanics is, as we argued in the last chapter, that of interpretation. What does the theory mean? What does quantum mechanics reveal about the nature of reality? It is when physicists have tried to address such questions that the reactions of Gell-Mann and Feynman have become commonplace.

There have been many attempts to make sense of quantum mechanics. The 'orthodox' interpretation, accepted by several generations of physicists, was formulated by Bohr and Heisenberg. It has come to be known as the Copenhagen interpretation, because Bohr was the director of the Institute of Theoretical Physics in Copenhagen. Heisenberg was Bohr's research assistant at Copenhagen from May 1926 to June 1927, and was thereafter a frequent visitor to Bohr's Institute.

The reality envisaged by Bohr and Heisenberg in their interpretation was not an objective, but a phenomenal one. It did not exist in the absence of observation. It was this, the Copenhagen interpretation, which has become increasingly influential over the course of the twentieth century. Einstein argued against the Copenhagen view for the last 30 years of his life. The disagreement between Einstein and Bohr on both the nature of quantum mechanics and its interpretation was to lead to one of the greatest debates ever in intellectual history.

In this chapter we outline the ideas that form the Copenhagen interpretation, together with some objections to it. We suggest that the Copenhagen interpretation is unacceptable, because it is based on the claim that quantum mechanics as it currently stands is a complete, final theory.

1900: Planck Ends Classical Physics by Announcing that Radiation is Quantised

The idea of the quantum was born in the first year of the twentieth century. On 14 December 1900, a German physicist, Max Planck, reluctantly announced that certain experimental results could best be understood if it was assumed that substances emit light only of certain energies.[3]

The origin of quantum theory lies in an everyday phenomenon. When heated, any piece of matter starts to glow, like a poker left in the fire. As the object gets hot, it gives off light. The colour of the light given off varies with temperature, going from red to yellow to white as the poker becomes progressively hotter.

The change in the colour of the light emitted represents a change in its wavelength and frequency. All bodies radiate, but the amount and the characteristics of their radiation depends only on two factors – the temperature of the body and the nature of its surface. An object does not need to glow to radiate. At room temperature, most radiation is in the infrared part of the electromagnetic spectrum. It is there, even if it is invisible.

The ability of a body to *radiate* is closely related to its ability to *absorb radiation*. This can be seen clearly if we consider a body at constant temperature. Such a body is said to exist in a state of equilibrium with its surroundings, since, for it to remain at constant temperature, it must absorb energy at the same rate that it emits it.

A *blackbody* is an idealisation conjured up by physicists to help them in their thinking about radiation. The term describes a body that is both a perfect absorber and a perfect emitter of radiation. A *blackbody* is so named because it absorbs all radiation that is incident upon it.

To approximate a *blackbody* in the laboratory, all that is required is a hollow box with a very small hole in one side. Any radiation which goes through the hole will enter the box, or 'cavity', where it remains trapped by reflection back and forth until it is absorbed. The cavity walls of the box will constantly emit and absorb radiation. That is why *blackbody* radiation sometimes goes under the name of cavity radiation.

When, in the nineteenth century, the more familiar phenomenon of heating a poker or any piece of metal was extended to a *blackbody*, no classical explanation could be offered that fitted the experimental facts. It was expected, that on heating a *blackbody*, the energy emitted at a given wavelength of radiation would increase steadily as the wavelength decreased. However, in experiments, it was discovered that a peak in energy actually occurred in the middle of a range of wavelengths. Nearly as surprising was a cut-off point at short wavelengths, even though the energy peak shifted toward the short wavelengths as the temperature increased.

The unexpected distribution of energy radiated from a *blackbody* commanded much scientific attention in the 1890s. Planck showed the way forward. By a combination of luck and insight, he argued that the distribution of energy could only be explained by postulating that energy was radiated in the form of discrete packets. He called these packets 'quanta'.

The idea that energy could only be emitted or absorbed in certain discrete amounts could not be fitted into the traditional Newtonian framework. Newton had believed that energy could have any value along a continuum. He had thought, too, that energy was transmitted not in pieces, but in an endless stream.

Planck's innovation, the quantum hypothesis, was such a departure that all physics which went before is now referred to as 'classical'. Planck grasped enough of the quantum's significance to feel exceedingly unhappy about his hypothesis. And even though the experimental facts could not be explained in any other way than by the quanta, Planck himself was only awarded the Nobel Prize for his work in 1919, nearly two decades after he had inaugurated the end of classical physics.[4] Such was the unease of his peers. In 1962, James Franck recalled watching Planck struggle

> to avoid quantum theory, [to see] whether he could not at least make the influence of quantum theory as little as it could possibly be... He was really a revolutionary against his own will... He finally came to the conclusion, 'it doesn't help. We have to live with quantum theory. And believe me it will expand. It will not only be optics. It will go in all fields. We have to live with it.'[5]

This was a fitting epitaph to a real, if reluctant, revolutionary.

Living with the Quantum (1): Discontinuity in the Atom

Physicists did have to 'live with' the quantum hypothesis; but they also used it. Niels Bohr was one of the first. He introduced the quantum idea of discontinuity directly into atomic theory. Bohr made use of Planck's hypothesis to develop a satisfactory model of atomic structure and stability.

The most widely accepted theory of the atom at the turn of the century was Thomson's 'plum pudding' model. J. J. Thomson, who had in 1897 discovered the electron, suggested that the atom was a sphere of positive charge in which was embedded, like plums in a pudding, negatively charged electrons.[6] In 1909, Ernest Rutherford supervised a series of experiments which subjected Thomson's pudding to a decisive test. Rutherford fired radioactive alpha particles – helium nuclei – at very thin sheets of gold leaf. On Thomson's model, Rutherford expected that the alpha particles would be only slightly deflected as they scraped by or travelled through the atoms of gold. However, while the majority of particles indeed emerged only slightly deflected, Rutherford found, to his complete amazement, that some were scattered in all directions; some even rebounded to their point of departure. 'It was', Rutherford said, 'as if you had fired a fifteen-inch shell at a sheet of tissue paper and it had come back and hit you.'[7]

In an attempt to explain his results, Rutherford revived a planetary model of the atom first put forward by Hantaro Nagaoka. Rutherford wrote to a colleague: 'I think that I can devise an atom much superior to J. J.'s'.[8] He was convinced that virtually all the mass of an atom was confined to a small, central, positively charged nucleus. Around this nucleus, negatively charged electrons orbited. The number of electrons had to coincide with the charge of the nucleus, since it was known that atoms carried no overall charge. In March 1911, Rutherford presented his results publicly for the first time.

The Rutherford atom, however, was no sooner announced

than it was beset by a problem. In classical physics, electrons orbiting a nucleus should radiate energy continuously and quickly spiral into the nucleus. Even stationary electrons would collapse, since opposite charges attract. Rutherford's atom proved an unstable structure. Atoms did not behave like this.

After a short time with J. J. Thomson's group at Cambridge, Niels Bohr joined Rutherford's team at Manchester in March 1912. Rutherford, essentially an experimentalist, immediately set Bohr the task of tackling the problem of atomic stability. Bohr believed that if the inevitable collapse suffered by the Rutherford atom was to be overcome, then a break with classical mechanics was required. The following year, Bohr presented, in three papers, the quantum theory of atomic structure.[9]

In the Bohr model, an electron still orbits around the atomic nucleus, but is no longer sent spiralling into it. Bohr's solution involved patching together classical physics with quantum theory. He took the orbiting electrons from Rutherford's model, and argued that they could not spiral into the nucleus by emitting radiation continuously, since radiation could only be emitted in quanta, *discontinuously*.

Moreover, Bohr's electrons were confined to certain 'stable' or 'stationary' orbits and no others. The permitted orbits were characterised by certain fixed amounts of energy. While an electron remained in one of these 'stable' orbits or 'energy levels', it could not absorb or emit radiation.[10] The electron could only absorb or emit a quantum of radiation, a photon, when it moved from one energy level to another. The frequency of the emitted or absorbed radiation was proportional to the difference between the energies of the levels concerned.

Bohr had constructed a model of the atom that tallied with Rutherford's experiment. He had, too, overcome the defects of Rutherford's model. Finally, he had provided, as Planck said during his Nobel Prize lecture, 'the long-sought key to the entrance-gate into the wonderland' of *spectroscopy*; a gate which, since the discovery of spectral analysis, had obstinately defied all efforts to breach it.[11]

Living with the Quantum (2): Bohr's Flawed Key to Spectroscopy

Spectroscopy had its origins in 1666, when Newton used a prism to break up white light into its constituent parts – the colours of the rainbow. It had been discovered in the mid-eighteenth century that different elements produced characteristic line spectra. These were used like fingerprints to identify the presence of a particular element. Nobody could explain why spectral lines appeared only at well-defined frequencies. Bohr's genius was to provide a link between line spectra and the atom.

An atom of a particular element has a characteristic number of electrons. For example, hydrogen has one. The electrons in the atom, according to Bohr, are allowed to undergo transitions only between fixed energy levels. When the electron of hydrogen 'jumps', it emits a photon that has an energy dependent upon the difference between the two energy levels.

It was already known that the energy of the absorbed or emitted photon is proportional to its frequency. In Bohr's conception, therefore, the spectra for hydrogen are a record of all the permitted 'jumps'. There is an exact match between the frequency of a spectral line, and the jump which corresponds to it.[12]

Bohr's explanation of line spectra allowed a picture to be built up of atoms with more complicated electron arrangements than hydrogen. It also cleared the way for an understanding of the periodic table based on the number of protons each element had.

However, problems beset Bohr's 'long-sought key to the entrance gate into the wonderland' of spectroscopy. When attempts were made to apply it to atoms with more than one electron, Bohr's model encountered inconsistencies. It was unable to account for the observed splitting of spectral lines; and in some cases it predicted too many lines. Quantum numbers – additional properties assigned to the atom – had to be introduced in an ad hoc fashion to make theory fit the experimental data.[13]

'Bohr had', notes Dugald Murdoch in his study of the Dane's philosophy, 'brought the notion of discontinuity, implicit in the quantum hypothesis, into particularly sharp focus: the discontinuous transitions between stationary states came to be known colloquially as "quantum jumps", a vivid metaphor for a kind of event that lay beyond the ken of classical physics.'[14] But such jumps were a problem not just for classical physics. They were also a problem for quantum theorists. Why, after all, couldn't the electrons exist anywhere between energy levels? And what happened to the electron during a jump?

Living with the Quantum (3): Discontinuity in Light – Einstein's Photon

Einstein was the first to learn to live with the quantum hypothesis, taking it more seriously than Planck. In 1905, the same year that he published his special theory of relativity, he also successfully explained the so-called photoelectric effect. Einstein did this on the basis that light is also quantized.

The photoelectric effect was first observed by Heinrich Hertz, during his work on the production of electromagnetic waves. Hertz noted that the space between two separated metallic plates, oppositely charged and contained in an evacuated tube, became a better conductor when illuminated by ultraviolet light. Others later showed that the effect is produced by electrons that are emitted from a metal surface when it is illuminated by light. The emitted electrons were referred to as *photoelectrons*.

One of the features Hertz's experiments revealed was that the maximum kinetic energy (energy due to motion) of the photoelectrons did not depend upon the intensity of the light beam striking the metal surface, but only on its frequency. Furthermore: the emission from any particular surface occurred only when the frequency of the beam of light that struck it was above a minimum value. For frequencies above this minimum, the emission of the photoelectrons occurred immediately, even when the intensity of light was very low.

These were all very surprising results. Einstein explained them by extending Planck's quantum hypothesis to light. For Einstein, light consisted of quanta, later called *photons*.[15] In the photoelectric effect, then, a beam of photons strikes the metal surface, whereupon the photons give electrons in the plate enough energy to escape. The fact that the energy given to an electron is dependent upon the frequency of the light used indicates that the energy of a photon is proportional to the frequency (or inversely proportional to the wavelength) of the light.

The greater the intensity of the incident beam of light, the greater the number of photons and consequently the greater the number of photoelectrons emitted. There is, however, a minimum energy required by an electron if it is to escape. This is why a minimum frequency has to be attained before emission occurs.[16]

In 1913, eight years after Einstein first put forward his hypothesis, the idea that light was composed of photons was still widely opposed. A strong reaction was evident even when four leading German physicists, including Planck, recommended Einstein for membership in the Prussian Academy of Sciences:

> In sum, one can say that there is hardly one among the great problems in which modern physics is so rich to which Einstein has not made a remarkable contribution. That he may sometimes have missed the target in his speculations as, for example, in his hypothesis of light-quanta, cannot really be held too much against him, for it is not possible to introduce really new ideas even in the most exact sciences without sometimes taking a risk.[17]

Similarly, as late as 1923, Bohr believed that the quantum hypothesis could not be considered a serious theory of light transmission. Light, he held, must be a wave-like phenomenon, since 'our description of radiation involves a large amount of physical experience involving optical apparatus including our eyes for the understanding of the working of which nothing seems satisfactory except the wave theory of light.'[18]

Einstein himself regarded light-quanta as a temporary expedient. In 1911, during the first Solvay conference of eminent physicists, held in Brussels, he said: 'I insist on the provisional character of this concept, which does not seem reconcilable with the experimentally verified consequences of the wave theory.'[19] Reluctance to accept the photon as a particle of light, then, was due in part to the confusion caused by Einstein's own stance.

However, in 1915, Einstein's work on the photoelectric effect was adequately tested by the American physicist Robert Millikan at the University of Chicago. It then started to gain wider support.[20] About his own position at the time, Millikan was to write later: 'I spent ten years of my life testing that 1905 equation of Einstein's and contrary to all my expectations, I was compelled to assert its unambiguous verification in spite of its unreasonableness'.[21] Finally, in 1922, Einstein received the Nobel Prize – not explicitly for relativity, but for his 'services to theoretical physics and especially for his discovery of the law of the photoelectric effect'.[22]

As we have seen, considerable disquiet surrounded the photon description of light. It was, after all, completely at odds with the classical view of light as a continuous wave phenomenon. Beneath the arguments about the nature of light lay the more profound fear that science was losing a definite picture of the way nature was. Did nature even have definite determinable properties any more?

To understand better the self-doubt which the quantum hypothesis prompted among the world's scientists, it is worth examining a little historical background to the discussion of the nature of light.

From Young's Waves to Compton's Quantised X-rays

Newton believed that light consisted of a stream of particles which he called 'corpuscles'. His assumption seemed to explain why light rays always appeared to travel in straight lines and why light bends when it passes from a less to a more dense medium – the phenomenon of refraction.

By 1730 Newton's corpuscular theory was accepted by everyone as the correct description of light. But in 1801, in one of the most famous experiments in the history of science, Thomas Young [1773–1829] seemed to show that light must in fact be a wave-like phenomenon. The modern version of Young's two-slits experiment runs as follows.

Two very narrow, close and parallel slits are illuminated by light from a single slit parallel to them, and placed directly in front of a monochromatic light source. A sodium discharge lamp is the source most frequently used. The two slits act as new sources of light once the sodium lamp is switched on. On a screen placed some distance in front of them, a pattern of dark and light bands known as 'fringes' can be seen.

An analogy helps here. Two stones are dropped simultaneously into a pond. Each produces small waves or ripples which spread out over the pond. Quite quickly, if the stones are dropped near each other, the ripples originating from one stone will encounter those from the other.

Where two troughs or two crests from the two stones meet each other, they only serve to reinforce the trough or crest. This is known as *constructive* interference. But where a trough meets a crest or vice versa, they cancel each other out. This is called *destructive* interference.

In Young's experiment, light waves originating from the two slits interfere with each other and strike the screen. The bright fringes are where constructive interference has occurred, while dark fringes indicate destructive interference. Only if light is a wave phenomenon can these results be understood.

Newton was so revered by the scientists of the day, and especially by the British, that Young's results were not immediately accepted. But Young was to have an ally – the French physicist Augustin Fresnel [1788–1827]. Wave theory was firmly established once Fresnel, whose experiments were more extensive than Young's, showed that it could explain a whole host of other optical phenomena.

By the mid-nineteenth century, Leon Foucault [1819–1868], another Frenchman, showed that the speed of light was greater in air than in water. This counted, once again, against Newton and in favour of Young. However, a new problem now loomed large. What were the mechanical properties of the 'luminiferous ether', the medium through which scientists believed light was propagated? Any successful theory would have to overcome immense difficulties. The ether had to be rigid enough to allow light to travel at great speed. But the ether also had to retain the ability to allow the planets to move through it unimpeded.

The first step scientists had to take was to discover the characteristics of light waves. That step was taken by James Clerk Maxwell [1831–1879]. Maxwell, the first professor of *experimental* physics at Cambridge University's Cavendish Laboratory, in fact made the most significant discovery in nineteenth-century *theoretical* physics.

Maxwell unified electricity and magnetism into electromagnetism. In the process he offered a new perspective on light. This he described as follows:

> The velocity of transverse undulations in our hypothetical medium, calculated from the electromagnetic experiments... agrees so exactly with the velocity of light calculated from the optical experiments... that we can scarcely avoid the inference that light consists in the transverse undulations of the same medium which is the cause of the electric and magnetic phenomena.[23]

Maxwell argued that light was an electromagnetic *wave* that travelled through the ether. It was Einstein, through his special theory of relativity, who finally showed that this medium was unnecessary.

Maxwell did not live long enough to see his work on electromagnetism confirmed by experiment. Aged 48, he died in the same year that Einstein was born – 1879. Within a decade of Maxwell's death, nevertheless, Heinrich Hertz produced electromagnetic waves and showed that they travelled at the same speed as light, as predicted by Maxwell's equations. In the course of his experiments, Hertz also observed the photoelectric effect.

And so we come full circle. For Einstein's explanation of that effect is reminiscent, not of the waves evoked by Young, but of Newton's 'corpuscles'.

By 1917 Einstein had overcome his reservations about light quanta. To his own satisfaction, he showed that photons possessed momentum – a *particle* characteristic. He wrote to his life-long friend Michele Besso: 'I do not doubt any more the reality of radiation quanta, although I still stand quite alone in this conviction'.[24] He then had to wait six years, until 1923, before irrefutable experimental evidence was forthcoming from the American physicist Arthur Compton.

Compton carried out experiments in which X-rays collided with free electrons. He found that, when an X-ray struck an electron, it was scattered from its original direction, and lost energy. At the same time, the electron gained energy and momentum, and recoiled. The scattered X-ray was found to have a lower energy than the original. The only manner in which Compton could explain his results was by considering X-rays as particles – as photons. The collision could be explained in terms similar to those required to explain a collision between two billiard balls.

An incoming X-ray could best be regarded as a photon, which, on striking the electron, transferred energy and momentum to it. After the collision, both the electron and X-ray photon are scattered at different angles. The analogy with the billiard balls breaks down slightly: the photon loses energy in the collision, so the frequency (and hence the wavelength) of the X-ray changes. In fact the scattered X-ray is a different X-ray from the incident one. This is called the *Compton Effect*. As a result of Maxwell's work, the Compton Effect can be extended to all electromagnetic radiation, including light.[25]

The Compton Effect 'created a sensation among physicists'.[26] After 1923, Einstein no longer stood alone, and the reality of the photon was readily embraced by nearly everyone. Arnold

Sommerfeld, a distinguished and leading figure in quantum research, was moved to comment: 'it is probably the most important discovery which could have been made in the current state of physics'.[27]

But why did the light-quanta hypothesis, first experimentally confirmed by Millikan, take eight years to become the fully-fledged photon and to be accepted? There are two specific reasons. First, as we have shown, the wave model was such an established part of scientific theory that it could not be easily overthrown. The general consensus was that experimentally-observed wave phenomena such as interference and diffraction ruled out any alternative approach.[28]

Second, the problems that faced the wave model, such as the photoelectric effect, were regarded as only temporary difficulties. It was believed that they would be overcome once the mechanisms whereby the electromagnetic waves interacted with matter were properly understood. As we have seen, even Einstein himself was susceptible to these views, as his comments at the first Solvay conference made clear.

More broadly, scientists were slow to accept the quantum hypothesis because they were disturbed by the prospect of losing a definite picture of reality. Indeed, it seemed that nature itself no longer had definite characteristics. The 1920s were to see these uncertainties magnified. Eventually, Bohr and Heisenberg proposed a solution which contained the emerging dualism in the physics of light. They created a unified picture of radiation – but only by jettisoning traditional ideas about the objective existence of the natural world.

Indeterminism in the Ascendant: De Broglie Establishes Wave-Particle Duality

After 1923 and Compton's work, it was accepted that Einstein's light-quanta provided an explanation of the photoelectric effect. But the photon description of light could not be invoked to explain the wave phenomena associated with light. Niels Bohr voiced just this objection to the light-quanta hypothesis at the third Solvay Conference, held in Brussels in October 1927:

> Such a concept seems, on the one hand, to offer the only possibility of accounting for the photoelectric effect, if we stick to the unrestricted applicability of the ideas of energy and momentum conservation. On the other hand, however, it presents apparently insurmountable difficulties from the point of view of the phenomenon of optical interference.[29]

Nor could a full explanation of Compton's experiment be provided without recourse to the wave model. The scattered X-ray was found to have a wavelength different from that of the original X-ray that had collided with the electron. Such a change in value of a property associated with waves was powerful evidence in their favour.

The *precise* change in wavelength, however, was inexplicable under the classical wave model. Only a quantised physics could account for it. On his own account, Einstein was well aware of the dilemma posed. In an article commenting on Compton's work, he eloquently expressed the widespread unhappiness, in the international physics community, at having to use 'two theories of light, both indispensable, and – as one must admit today despite 20 years of tremendous effort on the part of theoretical physicists – without any logical connection'.[30]

What Einstein meant was that both the continuous wave and discontinuous quantum descriptions were in some way valid; light had a dual, 'wave-particle' character. But if this concept took some thinking about, it was, prior to 1923, the discontinuity brought into physics by the quantum hypothesis that scientists found most difficult to accept. Planck had introduced that hypothesis only when all other possibilities were exhausted. Though he believed, as most physicists did, that radiation was both emitted and absorbed in discrete quanta of energy, he continued to maintain that this radiated energy travelled through space in a continuous waveform. Planck thought that the quantum description was only required in assessments of the exchange between matter and radiation; radiation, on its own in a vacuum, by contrast, propagated according to Maxwell's theory of electromagnetism.

The results of the experiments carried out by Compton made the view put forward by Planck inconsistent with scientific evidence. In 1923, the same year as Compton's experiment, a further blow was struck against Planck's interpretation by the French aristocrat Prince Louis de Broglie. In a series of papers, and in his doctoral dissertation the following year, de Broglie expounded the thesis that not only light, but *all* particles had a dual character.

De Broglie's examiners thought that his was a pretty piece of mathematics. But beyond that, they were unsure what to make of it. They agreed to send his dissertation to Einstein, who saw its significance immediately.[31] In 1929, de Broglie was awarded the Nobel Prize, by which time it had been shown that electrons did in fact exhibit characteristic wave properties. Today's electron microscopes exploit the wave nature of the electron.[32]

The problem facing physicists in 1924 was how to explain this wave-particle duality. To paraphrase and extend Einstein's point, there were now two theories, not only of light, but also of matter. Both were indispensable, yet the two were without any logical connection.

The Birth of Quantum Mechanics

The state of quantum theory at the beginning of 1925 has been described as 'a lamentable hodgepodge of hypotheses, principles, theorems and computational recipes'.[33] That situation was to be rapidly transformed by the emergence of *matrix mechanics* and of *wave mechanics*. These two theories were shown to be equivalent, and are today referred to as *quantum mechanics*. Quantum mechanics was to provide the logically consistent theoretical foundations that were lacking in what is now referred to as the 'old' (pre-1925) quantum theory.

In the September 1925 issue of *Zeitschrift für physik*, Heisenberg presented a remarkable article entitled 'On a Quantum-Theoretical Reinterpretation of Kinematics and

Mechanical Relations'.[34] The author's stated aim was ambitious: 'to establish a basis for theoretical quantum mechanics, founded exclusively on relationships between quantities which, in principle, are observable'.[35] Some 15 pages later, the aim was achieved, and Heisenberg had laid the foundations for a quantum mechanics of the atom and its interactions.

Years later, he lamented the resistance to quantum ideas which had preceded his work:

> When the physicists had met the existence of the quantum... people were able to feel that this was something important and very new. But they were not able to do the other step which would have been absolutely necessary to come further, and that is to throw away the old physics... Then what could the physicists do? They would of course try to use the old concepts and try to add if possible these new ideas in places where they found them necessary... Bohr, for instance, in his model of the atom, just used this idea of discontinuity to explain one essential fact, namely the stability of the atom. It took 12 years until one dared to really go away and push all the old concepts aside... You must remember that not only the tradition of 200 years of physics, but also all experience of planetary motions and everything else just proved that classical mechanics is right.[36]

Heisenberg was able to confront the weight of both tradition and experience.

The young Heisenberg's good fortune lay in beginning his studies at Munich University, under the guidance of Arnold Sommerfeld. Heisenberg then moved on to Göttingen, to become a research assistant under yet another leading quantum theorist, Max Born. Finally, as we have already mentioned, Heisenberg became Bohr's research assistant at the Institute of Theoretical Physics in Copenhagen. From the very start, he was exposed to the élite of scientists within the 'quantum triangle' of Munich, Göttingen and Copenhagen.[37]

Introduced to matrix algebra by Born, Heisenberg used it to explain the behaviour of electrons in atoms. In collaboration with Born and Pascual Jordan, he turned matrix mechanics into a mathematically self-consistent theory of the atomic domain. In particular, his 1925 paper successfully predicted those observed frequencies and intensities of line spectra which had remained unexplained by Bohr's work on the atom.

In 1925, most physicists were unfamiliar with matrix algebra. They were more inclined toward the wave mechanics of the Austrian Erwin Schrödinger, which appeared in print the following year.

A nod from Einstein in the direction of de Broglie's ideas on wave-particle duality set Schrödinger off. Schrödinger offered an analysis of quantum phenomena in terms closer to those of nineteenth-century physics than the highly abstract formulations of Heisenberg and Born.[38] He did not move in the same élite scientific circles as Heisenberg. As Professor of Theoretical Physics at the University of Zurich, he was removed from the major centres of quantum research and did not have any first-hand experience of the work being conducted in the 'quantum triangle'. However, in 1926 he published six papers outlining a distinctive formulation of quantum mechanics.[39]

Although the terms and concepts used by Schrödinger were more familiar than those of Heisenberg, the underlying ideas were just as revolutionary. Schrödinger introduced the notion of treating electrons as *standing waves*. This was a radical departure from thinking of electrons as particles.

Imagine holding a rope taut, with the other end fixed. With a flick of the wrist, it is possible to send a wave travelling down the length of the rope. This phenomenon is called wave propagation. A standing wave is formed when two waves, travelling in opposite directions, synchronise exactly. If the idea of standing waves is applied to Bohr's model of the atom, an electron's orbit can also be regarded as a standing wave. Each time an electron completes an orbit, it has travelled a distance into which only a certain number of whole standing waves can fit.

Schrödinger elaborated this idea into a fully-blown theory. In wave mechanics, the quantum conditions that were invoked *ad hoc* in the Bohr atom were found to arise in a natural way from fundamental postulates. They appeared as solutions of the wave equation postulated by Schrödinger to describe the wave properties of matter. Born believed that wave mechanics represented 'the deepest form of quantum laws'.[40]

Schrödinger believed he had explained why certain electron orbits were forbidden in the Bohr model. Only those orbits into which whole numbers of standing electron waves could be fitted were possible. He also believed that 'the lack of visualisability' in Heisenberg's matrix mechanics had been overcome. The Heisenberg approach left him 'discouraged, not to say repelled'.[41]

In return, Heisenberg was no more sympathetic: 'the more I think about the physical portion of the Schrödinger theory, the more repulsive I find it... What Schrödinger writes about the visualizability of his theory "is probably not quite right", in other words it's crap'.[42]

Heisenberg and Schrödinger put forward different formulations to account for the paradoxes thrown up by science in the period 1920–5.[43] But if we look at the *interpretation* of the new science, what is significant is that both formulations were widely taken to undermine the premises of objectivity and determinacy that are central to classical science.

The Oddness of the Quantum World (1): Statistics and Uncertainty

Quantum jumping remained as difficult to visualise in Schrödinger's model as in Heisenberg's. Schrödinger was driven to comment: 'If all this damned quantum jumping were really to stay, I should be sorry I ever got involved with quantum theory'.[44] Schrödinger's waves were regarded as abstract entities. They were not real waves in real space, but waves of probability.

Max Born was the leading advocate of this interpretation of Schrödinger's waves. Born was concerned to associate the 'wave function' with the existence of particles. He suggested that the square of the wave function gave the statistical probability of finding the associated particle at some particular point. His view was that it could never be known with certainty where a particle is, but that the square of the wave function at a particular point gave the statistical probability of finding the particle at that point.[45] Why? Because, taking the example of light, the intensity of light at a point is proportional to the square of its amplitude at that point. It can also be viewed as the number of photons at that point. Hence, the square of the amplitude of the light wave is proportional to the probability that a photon is present.

The introduction of statistics and probability into the realm of physics seemed to fit in well with Heisenberg's 1927 paper: 'On the Perceptual Content of Quantum Theoretical Kinematics and Mechanics'. This paper began one of the most important developments in twentieth-century physics.[46] In it, Heisenberg expounded his famous uncertainty principle.

One form of the uncertainty principle states that the position and momentum of a particle cannot be predicted simultaneously with a high degree of accuracy.[47] The more accurately one is measured, the less accurately the other can be predicted. The greater the precision and accuracy in determining the position of a particle, the greater the uncertainty introduced with respect to its momentum. There is a price to pay for the accuracy that is achieved.

The uncertainty principle appears to tell us that, in measuring one side of a pair of conjugate physical properties, the act of measurement causes fluctuations which are so unpredictable that the other side cannot be measured to an arbitrary degree of accuracy. From the wave-mechanical point of view, the principle says that a wave cannot simultaneously have both a well-defined wavelength (which depends on the wave being spread out) and a well-defined amplitude (which depends on the wave being squashed to reveal a peak at a particular point).

The Oddness of the Quantum World (2): Photons that 'Know'
When You're Looking at Them

Thomas Young's two-slit experiment offered experimental support for the wave theory of light. A reconsideration of the experiment taking account of modern quantum theory, on this occasion regarding light as photons, reveals just how odd the quantum world is.[48]

Suppose that the intensity of the incident beam of light is gradually reduced to such an extent that only one photon reaches the double-slit arrangement at any particular time. What happens next? The photon passes through one or other of two slits. Yet if a series of photons is sent through the slits, and their images recorded on a photographic film, the cumulative effect of 'corpuscular' photons hitting silver nitrate reveals... a wave-like interference pattern! This is an astonishing result.

If a photon remains a photon, it either goes through slit one or slit two: an interference pattern seems impossible. Yet if one slit is closed, the interference pattern disappears – exactly what would happen if light is a wave. The interference pattern cannot be recovered by superimposing the pattern obtained by first closing slit one and having slit two open and then opening the first and closing the second. Only if both slits are simultaneously open does the interference pattern reappear.

Altogether, it seems as if each photon individually takes into account whether only one or both slits are open. Yet how is this possible, if photons are indivisible particles? Either the photon goes through both slits at the same time, and is thus in two places at once. Or, each photon somehow cooperates in such a fashion as to build up an interference pattern. The question arises: how is this possible?

Accept, for a moment, that a physicist places detectors in front of both slits in order to determine which slit each photon is passing through on its way to the photographic film. The physicist could rig up an elaborate experimental arrangement, so that as soon as he or she determines through which slit the photon is about to pass, the other slit is closed.

The physicist now knows which slit the photon passes through. However, if this experiment is repeated a number of times and the film eventually examined, the pattern obtained is not one of wave interference, but that of particles striking the film. Again, it appears that the photon 'knows' that one slit has been closed, even though closure only occurs once the physicist has determined through which slit the photon is about to pass.

To understand what is happening, we have to take into account both Heisenberg's uncertainty principle and Born's probability waves. First, trying to determine which slit the photon is passing through, the physicist alters the photon's motion to such an extent that the interference pattern vanishes. Second, associated with each particle is an abstract probability wave. When the probability wave hits the double slit arrangement, part of the probability wave goes through slit one, a part through slit two. While a probability wave can be at both slits at once, a particle cannot. It is the probability wave emerging through the two slits that interferes to produce the pattern on the film. The intensity of the pattern is an indication of the probability of finding individual photons.

The quantum world is indeed an odd one. If a photon is left undisturbed by the physicist, then it behaves like a probability wave. However, if any attempt is made to look at a photon to establish through which slit it is about to pass, then it behaves like a particle.

The Copenhagen Interpretation: Retreat from Objective Reality

The uncertainty principle and Born's probabilistic way of coming to terms with the wave function together form the basis of the Copenhagen interpretation of quantum mechanics. This interpretation has become the orthodox one today – but only after a struggle.

At first, many believed that the uncertainty principle, as postulated by Heisenberg, was merely the result of technological limitations in the typical scientist's measuring equipment. If that equipment could be improved upon, it was argued, then uncertainty could be overcome.

In part, this misapprehension about measurement arose because of the way Heisenberg, in his original paper, had tried to draw out the significance of the uncertainty principle. Heisenberg suggested a thought-experiment, which is just an imaginary experiment using perfect equipment. His thought-experiment aspired to measure the position and momentum of an electron, using a microscope and light of very short wavelength.[49] Heisenberg explained that

> at the moment of position determination, when the light-quantum is diffracted by the electron, the momentum of the electron is changed discontinuously. The shorter the wavelength of the light, i.e. the more accurate the position measurement, the greater the change in momentum. At the moment the position of the electron is ascertained, its momentum can be known only within a magnitude that corresponds to this discontinuous change.[50]

It is clear from Heisenberg's argument that every attempt to measure a property at the quantum level inevitably results in an interaction with the measuring apparatus – and that this will lead to uncertainty.

Bohr later convinced Heisenberg that the uncertainty was not just a product of observation, but an actual feature of nature itself. It was not that the experimenter was too clumsy to measure the position and momentum at the same time. Rather, there was simply no such thing as a particle that possessed these two attributes simultaneously and to an arbitrary degree of accuracy. The uncertainty principle expressed an inherent limitation in nature. No improvement was possible, even with the most advanced measuring technologies.

Given this uncertainty, the concept of probability has to be used. In this interpretation, it is not possible to 'know' the position of a particle precisely. However, probabilities can be calculated that correspond to where that particle is likely to be under certain conditions. This indeterminacy, or uncertainty, is a complete departure from classical physics.

To explain the wave-particle duality, Bohr put forward his Principle of Complementarity.[51] The principle was introduced at a conference in Como, Italy, in September 1927. It was here that, as one commentator has described it, 'Quantum mechanics underwent an almost official inauguration'.[52] The Principle of Complementarity was to become a central pillar of the Copenhagen interpretation.

Bohr argued that the wave and particle properties of an entity, such as an electron or photon, were simply complementary. For him, 'evidence obtained under different conditions' could not be 'comprehended within a single picture', but rather had to be regarded as complementary in the sense that 'only the totality of the phenomena exhausts the possible information about the objects'.[53]

Bohr was adamant that no experimental situation would occur in which the distinct wave and particle aspects might conflict with each other. The two were mutually exclusive. Further, Bohr claimed that it was the nature of the experiment that was performed, on an electron for example, that revealed its particle or wave nature. This claim was an extension of the idea that a particle with a definite momentum does not exist until an experiment is performed to measure its momentum. Hence, in the Copenhagen interpretation, complementarity was used to underpin the uncertainty principle. If momentum is measured precisely in an experiment, then the position of the particle has to be uncertain. In the Copenhagen interpretation, momentum and the position of a particle are complementary.

For Bohr there was nothing odd in the photon two-slit experiment. Once an attempt was made to determine through which slit the photon passed, by using detectors, the experimental conditions had been altered. The wave aspect of the phenomena had been ignored; the attempt being made was to discern particle behaviour; so interference patterns were bound to disappear. Once the detectors were removed and no attempt made to discern the particle behaviour of the photon, the interference pattern would reappear.

A Subjective View of Nature

The Copenhagen interpretation suggests that *observation constructs reality*. Bohr wrote of 'fundamental limitations' within atomic physics, in the 'objective existence of phenomena independent of their means of observation'.[54]

The reality envisaged by Bohr was not an objective, but a phenomenal one. It did not exist in the absence of observation. Bohr did not actually deny the existence of an objective reality 'out there'; but he thought it meaningless to ask any questions about what this reality was. In Bohr's philosophy, the facts of measurement and observation must suffice. There is no point in asking what lies beyond the observation.

Einstein could not agree with Bohr. In 1954, a year before his death, he maintained:

> Like the moon has a definite position whether or not we look at the moon, the same must also hold for the atomic objects, as there is no sharp distinction possible between these and macroscopic objects. Observation cannot CREATE an element of reality like a position, there must be something contained in the complete description of physical reality which corresponds to the possibility of observing a position, already before the observation has been actually made.[55]

What, at root, was the difference between Bohr and critics like Einstein? In answering this question, we can clarify still further the competing philosophies which grew up around quantum mechanics.

In their collection of interviews *The Ghost In The Atom*, Davies and Brown provide an admirable summary of Bohr's approach. For Bohr, they point out, it was meaningless to ask what an electron 'really' is.

> Physics, he declared, tells us not what is, but what we can say to each other concerning the world. Specifically, if a

physicist carries out an experiment on a quantum system, provided a full specification of the experimental set-up is given, physics can then make a meaningful prediction about what he may observe, and thence communicate to his fellows in a well-understood language.[56]

Bohr's view has been well summarised by the physicist John Wheeler at Princeton University: 'no elementary phenomenon is a real phenomenon until it is an observed phenomenon'.[57]

Wheeler has illustrated what he means by telling a story. He once attended a dinner party which degenerated into a game of Twenty Questions. The aim of the game was to identify an object, selected by the other guests, through a series of 20 'yes' or 'no' answers to questions posed. Wheeler's fellow diners obviously knew him well enough not to think of anything at all, and instead decided simply to give answers consistent with those previously given. At the end of his series of questions, Wheeler believed that they had chosen 'cloud'.

Wheeler argues that 'in the game, no word is a word until that word is promoted to reality by the choice of questions asked and answers given'. This is the central point for Wheeler; but he also acknowledges the part played by the other guests. If they had responded differently, he would not have come up with 'cloud'. He believes his role in the game was the same as the role of the experimenter with electrons. Such an observer, Wheeler maintains, has a 'substantial influence on what will happen to the electron by the choice of experiments he will do on it, "questions he will put to nature"'. But the experimenter also knows that 'there is a certain unpredictability about what any given one of his measurements will disclose, about what "answers nature will give"'.[58]

Wheeler believes that it is only within the confines of a particular experimental situation that reality, phenomenal reality, can be specified. Moreover, he takes this belief to its logical conclusion: 'There is a sense in which what the observer will do in the future defines what happens in the past – even in a past so remote that life did not exist, and shows even more, that "observership" is a prerequisite for any meaningful version of "reality"'.[59]

What Wheeler means is that the observer literally creates the universe by his observations.[60] From Copenhagen, Denmark, to Princeton, New Jersey, the interpretation put on quantum mechanics was and remains a subjective one. Indeed, it often lapses into outright solipsism.

Copenhagen, Causality and Einstein's Response

As well as posing a profound challenge to existing views about the broad character of 'reality', the Copenhagen interpretation also challenged another cornerstone of the traditional approach to science: causality. For the Copenhagen school, the central role assigned to observation vitiated any and all attempts to unearth a regularity among or causal connections between phenomena. Indeed, Heisenberg generalised this rejection of causality from the quantum domain to the whole world of science: 'Because all experiments are subject to the laws of quantum mechanics, quantum mechanics definitely shows the invalidity of the causal laws'.[61]

While other interpretations of quantum mechanics exist, the Copenhagen interpretation remains the most influential. Some of the ideas surrounding this interpretation have found a wider resonance within science, and have been taken as starting points for research in other fields. This is especially so as far as the questioning of causal laws goes. As Paul Davies notes in a book concerned largely with chaos theory, modern physics has a strongly holistic character – a fact which 'is due in large part to the influence of quantum theory'.[62]

The question arises: is there a necessary connection between quantum mechanics as a mathematical formalism, and the conclusions drawn by Bohr and Heisenberg? Einstein, for one, remained unconvinced by the Copenhagen interpretation. In a letter to Schrödinger, he wrote of the Danish school: 'most of them simply do not see what sort of risky game they are playing with reality'.[63]

The question that lies at the heart of quantum mechanics is the question of interpretation.[64] The vast majority of physicists take the advice Feynman quoted at the start of this chapter: use quantum mechanics pragmatically, and leave philosophical questions to one side. Feynman, like Einstein before him, was referring to the disturbing consequences of the Copenhagen interpretation.

Copenhagen implied not just uncertainty and unpredictability, but a quixotic approach to the whole notion of an objective reality. Even Heisenberg had his doubts about the interpretation he helped found. In later life he looked back at a spell of intensive study with Bohr, when they covered all questions concerning the interpretation of quantum theory. He recalled that his discussion with Bohr 'finally led to a complete and, as many physicists believe, satisfactory clarification of the situation'. But it was not, Heisenberg continued, one which he could easily accept:

I remember discussions with Bohr which went through many hours till very late at night and ended almost in despair; and when at the end I went alone for a walk in the neighbouring park I repeated to myself again and again the question: Can nature possibly be as absurd as it seemed to us in these atomic experiments?[65]

If the Copenhagen interpretation made even Heisenberg despair at times, it was just unacceptable to Einstein and Schrödinger.

For Einstein and Schrödinger, Bohr and Heisenberg indeed played a 'risky game' with reality. From the fifth Solvay Conference held in Brussels in October 1927, until his death in 1955, Einstein was in outright dispute with the Copenhagen position. Toward the end of his life, he was to write of it: 'this theory reminds me a little of the system of delusions of an exceedingly intelligent paranoic, concocted of incoherent elements of thoughts'.[66] It was Einstein's belief that quantum mechanics was 'not yet the real thing'. There exists, Einstein wrote, 'a physical reality independent of substantiation and perception. It can be completely comprehended by a theoretical construction which describes phenomena in space and time. The laws of nature imply complete causality'.[67]

The Einstein-Bohr debate about how to interpret quantum mechanics had two distinct phases. The first was characterised by Einstein's attempts to show that Bohr's interpretation of the uncertainty relations was wrong. Einstein devised many thought-experiments which attempted to undermine them. In response, Bohr satisfied most scientists that Einstein could not carry through his desired aim each time.

The second, and more important, phase of the debate was characterised by Einstein's attempt to show that quantum mechanics was an incomplete theory. Here he developed one of his most enduring thought-experiments. Published in 1935 and written in conjunction with two younger colleagues, Boris Podolsky and Nathan Rosen, Einstein's paper was entitled: 'Can a Quantum Mechanical Description of Reality be Considered Complete?'.[68]

The 'EPR' experiment was formulated on the basis that either an independent reality did not exist, or quantum mechanics was an incomplete theory. Being a realist, Einstein firmly believed the latter. The thought-experiment addressed the problem of whether a particle could have both a definite momentum and a definite position. Einstein, Podolsky and Rosen devised a scheme in which it appeared that both these quantities could, in principle at least, be measured to *any* desired degree of accuracy – so contradicting the uncertainty principle.

Two particles, A and B, interact and then separate until they are quite far apart. It is possible to measure the momentum of particle A directly, and perform a calculation to determine the momentum of particle B. While it is not possible to know the position of particle A, because of the measurement performed on it, it is possible to determine the position of particle B directly. Consequently, the momentum and position of particle B can be determined, thereby circumventing Heisenberg and demonstrating the incompleteness of quantum mechanics.

In recent years, the EPR experiment has been realised in the laboratory.[69] For the present, the experiments appear to come down in favour of the Copenhagen interpretation and against Einstein's realism. But, accepting that the results of these experiments favour the Copenhagen position involves rejecting one of three premises on which Einstein based his position. These were that inductive logic is valid; that objective reality exists; and that it is impossible to travel faster than light.

The last premise is referred to as *locality*, or the principle of separation. It is actually a defence of causality. The third and last premise was very important for Einstein Podolsky and Rosen. Their original paper concluded by arguing that the Copenhagen interpretation 'makes the reality of [the momentum and position of B] depend upon the process of measurement carried out on the first system [A] which does not disturb the second system in any way'. No reasonable definition of reality could be expected to permit this.[70]

For the original EPR team, the idea that measurement could prompt simultaneous *action at a distance* was nonsensical. The recent experiments thought to favour Bohr over Einstein were only possible because of work carried out by John Bell in 1964. Bell took Einstein's three premises and derived from them correlations that should result between various physical attributes of a system as a whole. He expressed these correlations mathematically in what has become known as Bell's Theorem.[71] Nevertheless, Bell himself has suggested that there are problems with the laboratory experiments.[72]

The most popular interpretation of the experimental results is that while objective reality exists independent of observation, the speed of light can be exceeded. But difficulties then arise for the theory of relativity, since it is fundamental to this theory that nothing can travel faster than light.

Nobody is prepared to overthrow the theory of relativity. So the recent experiments which seem to vindicate Bohr in fact resolve nothing. To somebody who already believes in it, the results of these experiments only confirm Bohr's point of view. Conversely, to somebody who does not hold to the Copenhagen interpretation, the results do not matter. Indeed, in the light of the results to an experiment he himself had devised, results which many saw as 'proof' that Bohr was right, Bell himself was scathing about the Copenhagen approach:

> Well, it does not really explain things; in fact the founding fathers of quantum mechanics rather prided themselves on giving up the idea of explanation. They were very proud that they dealt only with phenomena: they refused to look behind the phenomena, regarding that as the price one had to pay for coming to terms with nature.[73]

However the results of the experiments based on Bell's theorem are interpreted, the original point of Einstein's thought experiment – to show that quantum mechanics is incomplete – still stands. The experiments only raise as many problems as they solve. Quantum mechanics remains, as Bell said, 'a dirty theory'.[74]

Copenhagen, Causality and Schrödinger's Response

Opponents of the Copenhagen interpretation sought to show that quantum theory was incomplete by routes other than that sketched out by the EPR team. Schrödinger in particular was concerned with the link between the macro and the micro worlds, and the relationship between this and the claim made by supporters of the Copenhagen interpretation that the act of observation played a central role in the creation of reality.

Schrödinger challenged the Copenhagen interpretation with his famous thought-experiment, known as 'Schrödinger's Cat'. Take a cat, said Schrödinger, and place it in a box together with a bottle of cyanide. Arrange things so that a hammer placed over the bottle will smash it when a single decay of a radioactive substance, also placed in the box, occurs. Now, according to Schrödinger and common sense, the cat is either dead or alive, depending on whether or not a radioactive decay has occurred. However, the realm of the subatomic, according to Bohr, is an Alice in Wonderland place. For Bohr and his followers would argue that, since only an act of observation decides if there has been a decay or not, then it is only this act of observation that determines whether the cat is dead or alive. If we do not look, the cat is consigned to Purgatory.

For Bohr, the act of observation did not necessarily mean the intervention of a human; it could have been the click of a counter or any such device. Some, most notably John von Neumann, have said that since macroscopic measuring devices are made up of atoms, then they should be subject to quantum effects. Von Neumann suggested that any measuring device employed to bring about an 'irreversible' act of measurement could only do so when it, too, underwent a measurement. The problem with this schema is that it demands infinite regress.

To highlight the problem further, Eugene Wigner replaced Schrödinger's Cat with a human. If after some time in the box, Wigner's friend is found to be alive and let out, the question can be put: 'Were you alive during the time you were enclosed in the box?' He will obviously say 'Yes'. Yet the Copenhagen view of quantum mechanics says that, before the box is opened, Wigner's friend must be in a live-dead state. However, he knows, and we now know, that during that time he was alive.

Schrödinger's thought-experiment showed that the Copenhagen interpretation jars with our commonsense understanding of the real world. Its purpose was to draw out the fantastic interpretation of the micro world proclaimed by Bohr – by magnifying that world up to the size of a cat.

The point raised by von Neumann was nevertheless useful. It served to highlight what we might call the 'boundary problem'. Where should scientists draw the line between the measuring apparatus, which is part of the macro world, and the object being measured, which is part of the micro world? The logic of von Neumann is that the Copenhagen interpretation arbitrarily gives a privileged position to the observer in the construction of reality. All matter is made up of atoms and therefore subject to the laws of quantum mechanics... and yet somehow the observer or measuring apparatus has a privileged position. No matter how much Bohr talked about the role of the observer, he never explained how such a beast could play this unique role.

In *Quantum Mechanics: Historical Contingency and the Copenhagen Hegemony*, James Cushing admirably sums up the question posed by Schrödinger. If we use quantum mechanics to describe the system, using the wave function for the whole system; cat and radioactive source together, what does the wave function represent? Cushing replies that there are two possible answers: (a) *our state* of knowledge (quantum mechanics is incomplete); or (b) the *actual state* of the system (there is a sudden change upon observation). He continues:

> If we choose (a) (which is what Schrödinger feels we must do intuitively), then quantum mechanics is incomplete (i.e., there are physically meaningful questions about the system that it cannot answer – *surely* the cat was *either* alive or dead *before* we looked). On the other hand, choice (b) saddles us with the measurement problem (and with a vengeance). The collapse of the wave function becomes an actual *physical* process that must be explained.[75]

Quantum mechanics does indeed have a great deal of explaining to do. As one physicist has recently argued: 'the "quantum measurement" paradox, so far from being a non-problem, is a sufficiently glaring indication of the inadequacy of quantum mechanics as a total world-view that it should motivate us actively to explore the likely direction in which it will break down'.[76]

How Bohr and Heisenberg 'Brainwashed' a Whole Generation of Physicists

After all is said and done, how should quantum mechanics be regarded, given its drastic implications concerning the nature of reality, and its many paradoxes? In our view, Einstein offered the best interpretation. His is an approach which recognises the major advances brought by the theory, while at the same time opposing the philosophical conclusions drawn by Bohr as simply unproven. Einstein held quantum mechanics to be an unfinished theory. It is, he said, 'an incomplete representation of real things, although it is the only one which can be built out of the fundamental concepts of force and material points (quantum corrections to classical mechanics)'.[77] At the same time, Einstein held to a view of quantum mechanics which he expressed in forthright style during his nomination of Heisenberg and Schrödinger for a Nobel Prize. 'I am convinced', he said, that this theory undoubtedly contains a part of the ultimate truth'.[78]

That much can certainly be said for quantum mechanics. Still, the assessment of Bohr and Heisenberg made by Gell-Mann, when writing of the project of understanding quantum mechanics, is in our opinion pertinent. Bohr and Heisenberg, says Gell-Mann, 'brainwashed a whole generation of physicists into thinking that the job was done 60 years ago'.[79]

Bohr and Heisenberg constructed a philosophical position based on their scientific outlook – a belief in the completeness and coherence of quantum mechanics. This led Bohr to hold that there was no quantum world:

> There is only an abstract quantum physical description. It is wrong to think that the task of physics is to find out how nature is. Physics concerns what we can say about nature.[80]

Einstein, on the other hand, believed in a causal, objective reality and based his understanding of quantum mechanics on this. As far as he was concerned, 'it is basic for physics that one assumes a real world existing independently from any act of perception.'[81] Consequently, he could not accept the Copenhagen interpretation.

Here is Heisenberg's response to the staunch stance taken by Einstein, Schrödinger and de Broglie:

> The Copenhagen interpretation of quantum mechanics has led physicists far away from the simplistic materialist

views that prevailed in the natural sciences of the nineteenth century... One attempt to counter the Copenhagen interpretation is an attempt to change the philosophy without changing the physics... a return to the ontology of materialism. They would prefer to go back to the idea of an objective world.[82]

Heisenberg rightly identified what Einstein and the others were up to. They *did* want 'to change the philosophy without changing the physics'. They accepted that quantum mechanics was the best theory available, although incomplete. It was the philosophy of the Copenhagen interpretation that they found most objectionable. They did, indeed, wish to return to the idea of an objective world.

Some have suggested that, given the Copenhagen interpretation, Einstein was wasting his time. Wolfgang Pauli wrote:

One should no more rack one's brain about the problem of whether something one cannot know anything about exists all the same, than about the ancient question of how many angels are able to sit on the point of a needle. But it seems to me that Einstein's questions are ultimately always of this kind.[83]

Other scientists continue to argue that Einstein was some kind of conservative, stuck in the past, who could not keep up with the latest developments. In fact, however, the reverse is the case.

While Heisenberg and Bohr claimed that their theory was complete, Einstein looked to the future. It was Heisenberg and Bohr, not Einstein, who had the closest affinity to the science of Newton. While the Copenhagen interpretation held that Newtonian concepts were inviolable for the macro world, Einstein wanted to develop a completely fresh approach – for both the macro and the micro worlds. As a minimum goal, he wanted the recognition that a new approach was necessary.

We have already cited Einstein's view that quantum mechanics is the best theory using the existing concepts.

Implicit within this was the need for *totally new concepts*. By contrast, Heisenberg argued that existing, Newtonian, concepts could not be challenged:

> The Copenhagen interpretation of quantum theory starts from a paradox. Any experiment in physics, whether it refers to the phenomenon of daily life or to atomic events, is to be described in the terms of classical physics. The concepts of classical physics form the language by which we describe the arrangement of our experiments and state the results. We cannot and should not replace these concepts by any others. Still, the application of these concepts is limited by the relations of uncertainty. We must keep in mind this limited range of applicability of the classical concepts while using them, but we cannot and should not try to improve them.[84]

This was Bohr's view also. At the 1927 Solvay conference, Bohr and Heisenberg made a joint declaration which included the following words:

> We regard quantum mechanics as a complete theory for which the fundamental physical and mathematical hypothesis are no longer susceptible of modification... Our fundamental hypothesis of essential indeterminism is in accord with experiment. The subsequent development of the theory of radiation will change nothing in this state of affairs.[85]

A more uncompromising defence of indeterminism would be hard to imagine.

Einstein highlighted the conservative character of the Bohr/Heisenberg approach with the following acid comment:

> The Heisenberg-Bohr tranquillizing philosophy – or religion? – is so delicately contrived that, for the time being, it provides a gentle pillow for the true believer from which

he cannot very easily be aroused. So let him lie there.[86]

Einstein's approach was audacious; Bohr and Heisenberg's was indeed sedative. Einstein was opposed to the 'risky game' put forward by Copenhagen, not because he was against risks, but precisely because he saw that Copenhagen meant anaesthetising scientists from ever taking them.

Chapter Four:
Chaos, Complexity and Control

Chaos theory and recent theories of complexity are linked by the fact that both are products or aspects of *non-linear systems*. A linear system is one in which different factors, or variables, interact in such a way that the overall outcome is the cumulative effect of relatively independent causal agents. A non-linear system does not have this property. Rather, a change in one variable affects the action of another, even if it has not changed itself. Note that non-linearity in a system does not by itself imply that the system will behave chaotically or in a complex manner. But if a system is behaving chaotically, or exhibits those particular forms of order which have been highlighted by the study of complexity, it is definitely a non-linear system.

There are a number of ways to analyse these different kinds of behaviour mathematically. The simplest is *dynamical systems theory*. Stephen Wolfram of Illinois University divides systems into four classes: stationary; periodic motion; chaotic; and complex.[1] Complex systems are to be found on the border between periodic and chaotic motion.

Our concern here is not with the mathematics of chaos and complexity, but with the interpretation of these two theories. We now give a short and non-mathematical introduction to the basic theories, before moving on to discuss their interpretation.

In the previous chapter, we highlighted the incomplete character of quantum mechanics. Nevertheless, quantum mechanics does capture a deep aspect of physical reality. With theories of chaos and complexity, on the other hand, there is much less of a consensus. Our own view is that, while some systems found in nature may be chaotic and/or complex, there is no evidence that nature *as a whole* is. In this chapter we raise, first, practical doubts about the universal claims made for the new theories. We then add our theoretical objections.

There is also no evidence that chaos or complexity theories capture the real essence of *social* systems. Yet ideologues of chaos and complexity believe that they can be extended, as theories, to cover this subject. They then proceed to identify limits to human abilities, which may not be as hard and fast as they suggest. In particular, they want to foreclose the ability of human beings fully to understand and control nature.

By contrast with the clear focus of the previous chapter on the subatomic realm, the claims made by theorists of chaos and complexity take this chapter into many different realms. However, we do highlight attempts to apply complexity theory to biology. It is in biology that the most serious work has been done to give complexity theory real meaning.

The Origins of Chaos Theory

In 1961, Edward Lorenz was merely one scientist among many modelling turbulence in dynamic systems. He worked at the Massachusetts Institute of Technology, studying long-range weather forecasting. A chance discovery led him to the conclusion that long-range weather forecasting was impossible. More significantly, it led to the rediscovery of chaos.

Lorenz had hoped to predict changes in the weather by simulating its behaviour on a computer. He was well aware that both his mathematical model and his machine were very primitive; but he imagined that better models and machines would enable a steady improvement in prediction over time, until eventually it would be possible to get machines to provide an accurate image of future weather patterns far into the future.

One day, Lorenz needed to re-examine some of his long-range predictions. Rather than run his programme from the beginning, he tried to save time by starting his sequence of computations half way. To do this, he fed in results for this half-way point – results which he already knew. Yet, to his amazement, the full-way pattern which the computer eventually outputted quickly diverged from its previous run; in a short period of time, it was vastly different. Why?

At the half-way point, Lorenz had only inputted the first three decimal places of his data. The discrepancy with his complete data was tiny, and he had assumed that it would make no difference. In fact, however, the computer rapidly magnified the difference, rather than averaging it out. Calculations showed that a more accurate computer would not solve the problem. So fast was any initial difference magnified that uncertainty, no matter how small, in the initial measurement of conditions made long-range forecasting redundant.

In his book, *The Essence of Chaos*, Lorenz takes sensitive dependence on initial conditions as the hallmark of chaotic behaviour: 'sensitive dependence can serve as an acceptable definition of chaos'.[2] Lorenz gave this definition an enduringly vivid and now popular image in a talk to the American Association for the Advancement of Science in 1972. The talk was titled: 'Predictability: Does the Flap of a Butterfly's Wings in Brazil Set off a Tornado in Texas?'.[3]

Feedback, Number Theory, Strange Attractors and Information Theory

At the heart of the new behaviour discovered by Lorenz is positive feedback. Here, the smallest of changes in one factor is amplified by another, second factor, in such a way as to bring changes to the first. Feedback in a guitar-amplifier system, for instance, consists of action, reaction and action again.

Strange as it might seem, the mathematics of folding and stretching pastry also illustrates the process of feedback quite nicely. Take a long thin stretch of pastry, and mark dots along its length. Then stretch it to twice its length and fold it in half, pressing it together again so that the length is the same as at the beginning. Repeat *ad infinitum*. Now, where will all the dots wind up, exactly? To this question, not even the most powerful computer in the world can give an answer in advance. As John Casti puts it:

> The stretching shows how points that start off nearby can lose sight of each other as the stretching continues, eventually failing to 'keep in touch'. Of course, the folding means that some points move closer together again. But it's impossible to know beforehand which points these will be. This is as good a definition as any I know of for what constitutes the essence of 'chaos'.[4]

Another example of chaotic behaviour can be found in the realm of number theory.

In mathematics, a *rational* number is one which can be expressed as a ratio of one whole number to another. By contrast, an *irrational* number is not expressible in such a form. The square root of two, for example, is an irrational number. Now: between any two rational numbers there is an irrational, and between any two irrationals there is also a rational. This intricate and never-ending nesting of rationals and irrationals can make for systems which are super-sensitive to their starting-points. Once a system is constructed that leads to one end-point if the starting-point is rational, and a different end-point if the starting-point is irrational, such a system will indeed be super-sensitive. The intricate shapes called Mandelbrot Sets, which have come to be synonymous with chaos for many people, work on a similar, simple principle. Yet they are thought to be the most complex mathematical objects known to humanity.

A final way of understanding chaos theory is to grasp it in relation to the three kinds of trajectory to which bounded systems were thought, before the discovery of chaos, to be exclusively drawn. These trajectories were known as *attractors*, and amounted to: point attractors – those steady states reached by systems as they run down; limit cycles – endless periodic motions, like that of a pendulum, for instance; and tori – the attractors associated with quasi-periodic motions, that is, motions which result from the superposition of different periodic motions (for example, the motion of a cat swung around in a spaceship orbiting the Moon as it orbits the Earth, which in turn circles around the sun).[5]

It was long assumed that the natural bounds on the motion of systems like weather, or folding pastry, limited their evolution to one of these three patterns. But chaos theory introduces a fourth, and rather bizarre attractor, appropriately called a *strange attractor*. The significance of this is that, in a chaotic system, super-sensitivity to initial conditions ensures that systems can take a range of different and unpredictable routes around and around its strange attractor. Moreover, although the attractor is bounded, the pattern it establishes for the system's behaviour *never* repeats itself.

More intriguingly still, while a line is one-dimensional, and a plane two-dimensional, a strange attractor has dimensions lying somewhere between one and two. It is more than a collection of lines, but less than a flat plane.

Unpredictability clearly lies at the heart of chaos theory. Stripped of inessentials, this key feature can be highlighted through the mathematics of information theory developed by Gregory Chaitin.

By studying the evolution of a system, we gain information about it. The art of forecasting is to gain information about the future of the system, or to express its evolution to its current state in a description shorter than the evolution was itself. With a simple system, like a pendulum rocking backwards and forwards, both procedures are easy. If we ignore the effect of friction, we know the entire future of the system as soon as we have observed one cycle – it will just repeat. Observing a few cycles will also allow the effect of friction to be taken into account, so that the pendulum's trajectory toward a stationary state can be mapped.

A system like a pendulum's motion is called *algorithmically compressible*, or *computationally reducible*. But chaotic systems do not have this property. We cannot predict their evolution. As Casti puts it:

Another aspect of the inherent unpredictability of chaotic processes is that their time evolution is what is sometimes called *computationally irreducible*. In other words, there is no faster way of finding out what such a process is going to do than just to turn it on and watch it unfold. In short, the system itself is its own fastest computer.[6]

Chaotic systems are systems totally unlike any others. This uniqueness can be studied using a range of different tools, most notably dynamical systems theory and information theory. Both bring out the central fact that Casti highlights: the unpredictability of chaos.

Are Nature and Society Chaotic?

Mathematically, the appeal of chaos theory lies in the ability of simple equations to generate complex and unpredictable behaviour. But does nature, not to speak of human society, really work this way? The idea that chaos undermines humanity's ability to know and plan follows from the claim that chaos is *more* than just mathematics. Most proponents of chaos theory have little doubt that their new theory can be applied to *all* of nature. In their book *The Matter Myth*, Paul Davies and John Gribbin claim that the whole universe is chaotic. If we humans cannot see this, it is because we are 'ignorant of the ultra-fine detail'. But we can hardly be blamed for this, for '*so is the universe itself*'.[7] The American journalist James Gleick, whose book *Chaos* became a bestseller, is equally sure of the universal truth of chaos:

> Simple systems give rise to complex behaviour. Complex systems give rise to simple behaviour. And, most important, the laws of complexity hold universally, caring not at all for the details of a system's constituent atoms.[8]

The sheer scope of action laid out now for chaos is remarkable. We hear only of new areas in which the theory applies – never of domains from which it is excluded.

The introduction to a standard undergraduate textbook on chaos follows a familiar argument:

> The remarkable fact that determinism does not imply either regular behaviour or predictability is having a major impact on many fields of science, engineering, and mathematics. The discovery of chaos changes our understanding of the foundations of physics, and has many practical applications as well. The subject sheds new light on the workings of lasers, fluids, mathematical structure and chemical reactions.[9]

In the same vein, Casti asserts that chaos theory can be comprehensively applied to the physical world:

> You might object that, although this kind of chaotic behaviour is displayed by the *mathematical* system, perhaps real life processes don't work that way at all. Comforting as such an argument might be, it just doesn't seem to hold up under detailed scrutiny... when it comes to assessing the predictability of any real-world process, we have to keep the possibility of chaos continually in mind.[10]

For Casti, 'any real-world process' may be susceptible to chaos.

In nearly all the texts which uphold chaos theory, logic runs in a familiar sequence. First, the theory is proclaimed to revolutionise our understanding of the physical world. Often, there then follows an exposition of the mathematics of chaos. Finally, we are given a few pages on how the application of the theory might be useful in particular domains.

In the early years of the study of chaos, this open-ended procedure was perhaps fair. But that it is still the path adopted by most books on chaos must raise doubts about the claims made for the universal application of the theory. Folding pastry does exhibit chaotic behaviour. But once we move beyond simple systems like this, difficulties begin.

Even practitioners of chaos are forced to make concessions about its ability to model real physical systems. Here is how Gleick sums up Lorenz's approach:

> Although his equations were gross parodies of the earth's weather, he had a *faith* that they captured the essence of the real atmosphere. That first day, he decided that long-term weather forecasting must be doomed.[11]

But is a leap of 'faith' good science? Ruelle, who defends Lorenz's conclusions, is well aware of the methods by which they were arrived at:

> By a crude approximation, Ed Lorenz replaced the correct time evolution in infinite dimensions by a time evolution in three dimensions, which he could study on a computer... The Lorenz time evolution is not a realistic description of atmospheric convection, but its study nevertheless gave a very strong argument in favour of unpredictability of the motions of the atmosphere.[12]

Approximate models, like faith, are all very well. But there is a need to be very cautious: unpredictability in physical systems only follows if they really are chaotic, not just approximately so.

In the rush to apply non-linear mathematics to all physical systems, it is all too simple to forget the ABCs of the scientific method. Peter Coveney and Roger Highfield, in their book *The Arrow of Time*, do just this. To illustrate, we briefly consider these authors, beginning with their study of animal population change.

Abstraction, Pattern Recognition and the Facts

Models of population development are notoriously difficult, because of the wide variety of interdependent relations between different predators and their different preys. Changes in food supply also complicate matters. Still, Coveney and Highfield skate past the difficulties by using chaos theory. Writing of lynxes versus hares, they contend:

> Thanks to non-linear dynamics, an alternative explanation has been proposed that can be couched in terms of properties of the lynx-hare populations alone, without the need for mishaps in the snowshoe hare's food supply, weather fluctuations, disease or other external factors. In a non-linear dynamical system, the irregularities might owe their existence to chaos.[13]

Coveney and Highfield strip out the 'mishaps' of the world, go on to deploy the subjunctive 'might', and so squeeze the irregularities of lynx/hare relations into a chaotic framework. Essential aspects of real life are set aside in the construction of the very mathematical model that is supposed to explain them.

Of course, in building models, scientists necessarily abstract out from certain aspects of a problem. However, if the aspects dispensed with are actually central to the problem, models built will not even be approximations. The procedure adopted by Coveney and Highfield is used by many others in the chaos field: physical systems are modelled using chaotic *mathematics*, and thus held to be chaotic. But the models built often display myopia about the real aspects of the systems being studied.

Another approach to examining whether a natural system is behaving chaotically is to look for some of the tell-tale signs of mathematical chaos in its patterns. Studying the evolution of a system in time might reveal whether a strange attractor is present, for example. The enterprise is a useful one: if there are signs of chaos in the data, it is a sensible project to construct a chaotic model of the system.

As Coveney and Highfield themselves admit, however, claims for chaos theory have to be qualified in the light of the hard facts. Signs of chaos are by no means evident from the data which has been collected on different systems:

> We must distinguish between deterministic chaos –

intrinsic to the system – and randomness caused by a cacophony of external influences. There are good tests to differentiate between them but they are not straightforward to put into practice. For without detailed data, existing methods for detecting the hallmark of chaos – such as the fractal dimensions of purported strange attractors – cannot be relied upon. No one has yet proved that even the most well documented examples of these attractors fulfill rigorous mathematical definitions.[14]

Coveney and Highfield go on to concede an important point. The more variables that are taken into account in modelling a system, and thus the more realistic the model is made, the more difficult it becomes to unearth real chaos. About the weather, the authors write: 'in the face of the hyperbole, it is usually forgotten that if one adds further variables to Lorenz's equations in an attempt to make the picture more realistic, chaos becomes harder, not easier, to find.'[15]

As with natural systems, so with human ones. Despite a significant research effort, David Parker and Ralph Stacey have found that there is no evidence for chaos in real economic systems in their study for the Institute of Economic Affairs.[16]

The practical difficulty finding chaos in real data, and of accurately modelling systems using chaos theory, means that the mathematics of the subject recruits more bright minds than does the attempt to *apply* it. Lorenz quickly quit meteorological research for the pure maths of chaos. Benoit Mandelbrot, father of fractals, followed a similar route. As Edgar Peters puts it:

At the end of one entirely theoretical paper with no empirical proof to back up its arguments, Mandelbrot (1972) promised to publish statistical results, but he never did. Mandelbrot largely left economics to go on to broader work, developing fractal geometry.[17]

Robert Savit, professor of physics at the University of Michigan, is another exponent of chaos theory. But in relation to the behaviour of stock markets, he admits that maybe linear equations, together with a lot of 'noise' (a 'cacophony of external influences' as Coveney and Highfield call it), form a better approximation of events than chaos theory:

> Because real economies are so much more complex [than chaos-based models], it is possible that the chaotic effects we see in the games may simply not be present in a real economy. The real system may have so many different things going on that all interesting deterministic effects are just averaged away. We may find ourselves back in a situation in which the economy can be described by linear processes with a lot of noise thrown in.[18]

Our survey of some of the current literature on chaos theory has brought out three points. First, chaotic models often disregard key aspects of the reality they are supposed to describe. Second, in data on real systems, genuine signs of chaos have proved very elusive. Finally, linear models plus 'noise' are a credible alternative to chaos theories. Altogether, then, there is little hard evidence that chaos theory provides a model for the workings of natural or social systems.

Two Theoretical Doubts about the Universal Claims Made for Chaos Theory

There are two powerful theoretical objections that can be raised against the notion that chaos is a universal theory of nature.

First, despite his popular books in favour of chaos as a universal theory, Paul Davies himself has contested such claims. At a conference held in 1989 to study order and complexity, Davies pointed out that the very fact that humans exist and can understand the large-scale functioning of the universe implies that the universe as a whole must be in some way algorithmically compressible. In other words, it cannot be totally chaotic. Davies says that he does not know why this is the case, but it obviously is:

Our mental model of the world is itself an algorithmic compression. If the world were not compressible in this way, there could be no cognition. So the very fact that we exist as observers already constrains the universe to have the property of algorithmic compressibility. Of course, this anthropic reasoning does not constitute an *explanation* of why the universe is compressible; it merely tells us that we could not be around to debate the issue were it not so...

There is a wide class of physical systems, the so-called chaotic ones, which are *not* algorithmically compressible. One can imagine a universe in which there are no regularities at all, only chaos. The fact that there is cosmos rather than chaos is the starting point of science.[19]

Savit, as we have observed, thinks that stock markets might at a fundamental level be linear rather than non-linear. For his part, Davies goes on to speculate that linearity might be the only way to explain the comprehensibility of the universe:

The existence of non-chaotic dynamical regimes is a profound fact about nature, and one can ask for an explanation of this fact. I don't know whether we will ever have such an explanation, or what sort of explanation it might be. But one possible strand of reasoning might run thus: The non-chaotic nature of many systems seems to hinge on their approximate linearity and this, in turn, depends on the smallness of certain coupling constants, radiative corrections, etc. At present we do not have a theory of coupling constants, but one may emerge from attempts at grand, or super, unification. If a satisfactory unified theory is found for which these constants are fixed, then that will constitute a partial answer as to why chaos is kept at bay.[20]

We need not follow Davies down the rather speculative road he maps out to appreciate his underlying argument. Algorithmic compressibility is a brute fact of the natural world. In turn, that means that chaos cannot be a theory of the whole universe.

Our second theoretical doubt about the universality of chaos concerns the quantum domain. There, it is possible that chaos is damped out of existence. But if, as we have argued, quantum mechanics does, in powerful style, capture a fundamental aspect of nature, this damping effect might be as fundamental – maybe more fundamental – than chaos. Were that to be the case, damping would throw doubt on the ability of chaos to model *any* physical system.[21]

If chaos does *not* provide a universal theory of nature, that might be good news. The fact that *some* systems are chaotic might be something that humanity can use to its advantage. An understanding of chaotic systems could then enhance, rather than diminish, human powers. Once the universal claims for the theory are abolished, perhaps, chaos might turn out to be a powerful force in nature which humanity can harness.[22]

Ilya Prigogine and the Theory of Complexity

Exploring Complexity (1989) is the title of a study by Ilya Prigogine. It is a very appropriate title for study into what is a fast-growing subject and one that is still only vaguely mapped out. As we indicated in Chapter Two, many scientists in the field hope that complexity will one day enable a very wide range of phenomena – from ecosystems to economies, from the development of human embryos to the development of ideas – to be studied using a similar framework.

The central issue of interest in complexity theory is the study of *order*: its origin, character, and stability or instability. The different ideas in complexity theory can be illustrated by the approach taken to the study of order, and by the way in which this contrasts with the traditional approach taken by theoretical physics to the study of natural laws.

It is not obvious why order should exist. As Davies has noted:

> Given the limitless variety of ways in which matter and energy can arrange themselves, almost all of which would be 'random', the fact that the physical world is a coherent

collection of mutually tolerant, quasi-stable entities is surely a key scientific fact in need of explanation.[23]

Contemporary science suggests that total dis-order is more likely than order. In particular, the second law of thermodynamics, which argues that disorder grows with time, has stacked the cosmic cards against order.

In Prigogine's theory, the phenomenon of feedback is as central as it is to chaos theory. The reason for this is that Prigogine is interested in non-linear systems. Prigogine argues that non-linear systems can generate a novel kind of order: an order which, because of its origins in non-linear feedback, is delicately poised and impossible to predict and control. Moreover, Prigogine believes that feedback in natural systems is only a manifestation of something deeper.

Natural systems, Prigogine holds, develop through a process which he calls 'self-organisation'. It is this idea that links his scientific views to his philosophical ones. In one of his popular works, Prigogine says that science is

heading towards a new synthesis, a new naturalism. Perhaps we will eventually be able to combine the western tradition, with its emphasis on experimentation and quantitative formulations, with a tradition such as the Chinese one, with its view of a spontaneous, self-organising world.[24]

Prigogine's view is that, in far-from-equilibrium conditions, systems have the capacity for self-organisation, and that this is in fact a general principle, a general law of nature.

Prigogine's starting-point is *irreversibility*. He believes that not only systems, but natural laws themselves, evolve with time. The universe is irreversible. The classical framework of equations is all very well, as are the equations of nineteenth-century thermodynamics – in which time can theoretically be run backwards simply by substituting minus time for time throughout. But for Prigogine, irreversibility is a fact of nature.

Because of the super-sensitivity of the natural world,

Prigogine argues, the whole of nature is forced to pass through points of *bifurcation*: points from which there is no return. This makes forward movement into the future, rather than backward movement into the past, built into the very way in which nature is constructed. Nature has an 'arrow of time'.

An example illustrates the novelty of Prigogine's view. People are fairly used to seeing tea cups smash. They do not see broken fragments gathering themselves into tea cups again. If a film showed this, they would immediately protest that it was being run backwards. But Prigogine wants to say more than this. Mainstream science would concede that while it is *possible* for cups spontaneously to regenerate, such an event is just so unlikely as to be not worth considering. In other words, the irreversibility of natural processes is of a statistical origin, whereas natural law itself is reversible. Prigogine disagrees. He says that the recomposition of a tea cup is logically and thus totally impossible, because natural law is irreversible.

What help is this in understanding how order can emerge from disorder? Prigogine argues that, in far-from-equilibrium conditions, systems which are open to their surrounding environment exhibit *autocatalysis*. That is, they generate a system of internally reinforcing movements which lead to the build-up of order. Each bifurcation opens the possibility of greater order. The order is delicately poised because each bifurcation event is a chaotic phenomenon.

Prigogine's theory is built upon the breakthrough he made in 1967 when he formulated the idea of *dissipative systems*. These are systems that are able to generate order by dissipating disorder into their surrounding environment. In this way, order is generated in a particular area without violating the second law of thermodynamics overall.

At the time Prigogine discovered dissipative systems, it was well-known that *life* involves the creation of local regions of order. Prigogine's achievement was to develop the idea that such local order was in fact a general property of natural law, which simply found one expression in life. As he puts it in *Exploring Complexity*:

We find a deep unity among diverse systems, and this allows us to assert that the passage toward complexity is intimately related to the *bifurcation* of new branches of solutions following the *instability* of the reference state, caused by the non- linearities and the constraints acting on an open system.[25]

For Prigogine, order is both producer and product of the 'innovation and diversification' exhibited by all natural systems.[26] Prigogine goes further: 'irreversibility has a *constructive* role. It makes form. It makes human beings.'[27] Ultimately Prigogine leads us to a new view of matter. It is a view in which, he argues,

Matter is no longer the passive substance described in the mechanistic world view but is associated with spontaneous activity. This change is so profound... that we can really speak about a new dialogue of man and nature.[28]

Matter, in Prigogine's schema, has a life of its own.

Prigogine's difference with classical science is a stark one. For Darwinism, the ability of DNA to synthesise and create order is a brute fact, and not an expression of something inevitable. And for many prime movers in nineteenth-century thermodynamics, such as Ludwig Boltzman, the second law of thermodynamics was a statistical outcome rather than an essential law of nature, as it is in Prigogine's view.[29]

Reductionism, Emergence, Adaptation and Chaos

For many years, Prigogine was one of a very small number of theorists interested in complexity and order. Since the mid-1980s, however, there has been an explosion of interest in the field. Complexity theory has become a subject in its own right; and a number of centres are dedicated to studying it and its implications.

The easiest way to summarise the different approaches to the subject is to continue to pursue the discussion of order. This is a subject on which there is not a consensus. One thing is agreed, however: the view of order held by traditional science is insufficient, perhaps even wrong. So what is this traditional view?

Steven Weinberg presents a forceful exposition of it in his book *Dreams of a Final Theory*. According to Weinberg, complex systems, including humanity, can be understood using the idea that nature is the product of timeless fundamental laws interacting, in time, with accidents. In this view, order is a 'frozen accident'. The idea was given a poetic formulation by Monod, when he described life as 'chance caught on the wing'.

Of the human mind, Weinberg writes: 'objective correlates to consciousness can be studied by the methods of science and will eventually be explained in terms of the physics and chemistry of the brain'.[30] In turn, physics and chemistry will themselves, in time, be explained in terms of particle physics. Summed up, Weinberg's is the famous programme of reductionism. Indeed, the third chapter in *Dreams of a Final Theory* is titled 'Two Cheers for Reductionism'.

By contrast with traditional reductionism, complexity theory is concerned with 'emergence'. The basic idea is that, at each level of nature, new laws and phenomena emerge that are more than just 'chance caught on the wing'. Put at its strongest, the idea suggests that, at each new level, laws which apply at a lower level may be largely irrelevant. Thus, for Jack Cohen and Ian Stewart, in their *The Collapse of Chaos*, particle accelerators may tell us about the constituents of atoms, but may otherwise represent a waste of money – because they will tell us very little about the larger-scale workings of nature.

A less extreme view is put forward in *The Quark and the Jaguar* by Murray Gell-Mann. Without belittling the study of nature at all its levels, Gell-Mann diverges sharply from the view that order is simply a 'frozen accident'. In his book, he distinguishes between *complex systems* and *complex adaptive systems*. An example of the former is a galaxy; an example of the latter is life, or human consciousness. The difference is crucial. While the former can be understood using conventional physics, the latter are something altogether different. They are complex systems, but they are much more than that – they have the capacity to compress information; to learn from their environment; and, through this process, to become more complex, through self-organisation.

Again, like Prigogine, Gell-Mann goes further. For him, systems which adapt do so by selecting for increasing complexity:

> As time goes on, more and more frozen accidents, operating in conjunction with the fundamental laws, have produced regularities. Hence, complex systems of higher and higher complexity tend to emerge with the passage of time through self-organisation, even in the case of nonadaptive systems like galaxies, stars, and planets... In the case of complex adaptive systems, that tendency may be significantly strengthened by selection pressures that favour complexity.[31]

It is not clear from Gell-Mann's book just how much he draws a contrast between the two different kinds of complexity he describes. Still, his argument is that complex adaptive systems are neither totally random in their movement, nor totally ordered. It is their position mid-way between these two poles that gives complex adaptive systems the property of being adaptive.

Gell-Mann's idea has been given a rigorous treatment by Stephen Wolfram, who has his own centre for the study of complex systems at the University of Illinois. In the language of dynamical systems, Wolfram contends, complexity is a class of behaviour on the cusp between periodic attractors and the strange attractors described by chaos theory.

Wolfram's perspective is very close to the notion of complexity and order developed at the Santa Fe Institute – the notion of there being an 'edge of chaos' between stability and chaos. In his popular exposition of complexity theory, *Complexity: The Emerging Science at the Edge of Order and Chaos*

(1992), Mitchell Waldrop summarises the views of Christopher Langton, a founder of the Institute. He has, Waldrop suggests,

> [an] irresistible vision of life as eternally trying to keep its balance on the edge of chaos, always in danger of falling off into too much order on the one side, and too much chaos on the other. Maybe that's what evolution is, he thought: just a process of life's learning how to seize control of more and more of its own parameters, so that it has a better and better chance to stay balanced on the edge.[32]

Like Prigogine, Langton and others at the Santa Fe Institute see order as both an integral part of natural law, and also as a delicately poised thing. Liable to react in unpredictable ways to shocks, order might always jump into an entirely different state.

Complexity theorists believe that their theory can be applied to any system that exhibits order – from the origins of life, to the stockmarket, to human consciousness. Some work has been done in applying complexity to economics; some is just beginning on its applicability to the Earth's eco-systems. Most work, however, has focused on applying it to biology.

Has that work in biology been successful? Is complexity theory 'mathematically interesting but biologically trivial', as Oxford ecologist Robert May believes, or is there more to it than that?[33] An examination highlights some of the problems that surround the idea that complexity theory provides a new universal theory of nature and society.

Stuart Kauffman, Order and the Meaning of Life

The Origins of Order (1993) is Stuart Kauffman's epic reinterpretation of biology. Kauffman reinterprets three great themes: evolution; the origins of life; and the development of individual organisms. His is the most ambitious attempt yet to show how complexity theory leads to a new outlook on old problems. Kauffman seeks to show how elements of complexity theory, and in particular the 'edge of chaos' and 'emergence', are directly relevant to a study of a real biological system. Kauffman's starting-point is that evolution is

> not just 'chance caught on the wing'. It is not just a tinkering of the ad hoc, of bricolage, of contraption. It is emergent order honoured and honed by selection.[34]

In opposition to the usual Darwinian view that organisms evolve through chance mutation and natural selection, Kauffman argues that there are a number of preferred forms, or types, of organism. These forms, or types, are what natural selection is largely constrained to bring about. 'Selection', Kauffman maintains, 'has not struggled alone these many eons'.[35]

Mainstream biology holds that the origin of life on earth, some 3.8 billion years ago, was an accident. There are a number of competing hypotheses about how life moved from the most primitive beginnings to being a system based on DNA; but they are all based on the idea that accident played a large part at those beginnings. Here, however, Kauffman disagrees. Life, he says, is an expression of ordering principles in non-living matter. *Life is to be expected given complexity theory*:

> For all the known organisms on this branching river we call life, biology should aim ultimately to account for those essential features which we would expect to find in any recurrence of such a river. To suppose, as I do, that such an intellectual task may one day be achieved is, among other things, to suspect with quiet passion that below the particular teeming molecular traffic in each cell lie fundamental principles of order any life would re-express.[36]

Were the universe to happen all over again, Kauffman suggests, life would occur again.

In the third and final section of his book, Kauffman devotes himself to the development of individual organisms from fertilisation to maturity, a process known as *ontogeny*. Here he begins with what he calls 'perhaps the most critical single fact about ontogeny': the fact that, in any organism, the set of genetic instructions in each cell is identical in all cells.[37]

That fact means that each cell carries, within it, the genetic instructions needed to make the whole organism. Mainstream biology holds that such a rule just happens to be the way evolution turned out. But Kauffman thinks differently. According to him, ontogeny is 'an overwhelming example of self-organisation'.[38]

The big issue in ontogeny is how overall patterns of development are regulated – how cells know where to go, when to divide and how to grow. The traditional Darwinian perspective, developed in large part by Monod, emphasises the way in which some genes regulate others. Kauffman concurs, but goes on:

The order seen in ontogeny, I shall suggest, is just that which arises spontaneously in the powerfully ordered regime found in parallel-processing networks. Selection, I shall further suggest, by achieving genomic systems in the ordered regime near the boundary of chaos, is likely to have optimised the capacity of such systems to perform complex gene-coordinating tasks and evolve effectively.[39]

The regulation of some genes by others is, for Kauffman, itself a symptom of a deeper order.

Kauffman does not pose his work as an alternative to Darwinism. He sees it as a development of classical theory – an attempt to situate Darwinism in a deeper context. He does, however, draw fairly explicitly upon three traditions which are hostile to Darwinism.

First, Kauffman's emphasis on form as a constraining factor in natural selection is quite similar to themes put forward by the brilliant, if eclectic, anti-Darwinian D'arcy Thompson in his book *On Growth and Form* (1917). Second, Kauffman's emphasis on life as an expression of a process of increased complexity in nature is notable for the similarities it has with the writings of Jean-Baptiste de Lamarck [1744–1829].

Lamarck is best known for his idea that organisms change

through parents passing on acquired characteristics to their offspring. This, however, is only a secondary aspect of his theory. Lamarck's central idea was that organisms contain an inherent drive towards greater complexity and perfection.

Third, Kauffman's idea that the whole of life on earth is drawn to a more-or-less ideal state called the 'edge of chaos' bears more than a passing resemblance to the famous *Gaia hypothesis*. Kauffman himself says: 'suppose it's really true that coevolving, complex systems get themselves to the edge of chaos. Well, that's very Gaia-like'.[40]

Gaia is the idea that life as a whole seeks to maintain the conditions suitable for the maintenance of life. As a 'holistic' theory, it runs counter to Darwin's insistence that no species can evolve for the good of another. Gaia also abolishes natural selection altogether: if life on earth can be considered as a single organism, there can be no competition with other life forms – the scramble for reproductive success through selection is no more.

We have seen that Kauffman's professed Darwinism is, at best, debatable. We now address this issue in detail.

Natural Selection Downgraded

We discuss Kauffman's basic ideas by examining his treatment of evolution and form. Kauffman uses two key concepts in his model of evolution. These are the 'NK' idea of *fitness*, and the idea of a *fitness landscape* across which organisms roam.

The fitness of an organism – its capacity to thrive and prosper in a given environment – is, for Kauffman, the result of the interaction of N and K, where N is the number of genes in the organism, and K is the average number of genes which affect the functioning of each particular gene. The overall effect is computed by a subtle averaging process:

> The NK model consists of N sites, each with A alternative states, or alleles. Each site receives epistatic inputs from K other sites chosen among the N. The fitness contribution of each site for each of the A^{k+1} combinations of alleles of itself

and the K other sites is assigned at random from the uniform interval between 0.0 and 1.0. Thus given any genotype for the N sites, the fitness contribution of each site in the context of its K inputs can be calculated. The fitness of the entire genotype is the mean of the fitness contribution of its N sites.[41]

Kauffman's central idea is that, for different values of N and K in an organism, a differently-shaped fitness landscape will emerge, given any specific environment surrounding the organism.

The shape of the fitness landscape will determine the *attainable* level of fitness for an organism. In Kauffman's world view, then, natural selection is consigned to a secondary role. Kauffman argues that the interaction of N and K gives fitness landscapes definite properties – properties independent of natural selection. *Order is generated without selection.* More than this, Kauffman argues that certain values of K will lead to ideal fitness landscapes.

Taking one more step, Kauffman contends that natural selection will merely realise those values of K which lead to these ideal landscapes. In any particular environment, those values of K will generate an ideally-shaped landscape.

Clearly, for different kinds of system, with different rates of adaptation and other variable factors, different values of K will turn out to be optimal. Kauffman claims that, if we look at eco-systems, K-values between eight and 10 will generate optimum landscapes.[42]

Organisms blessed with this number of genes affecting each gene not only achieve a high degree of fitness; they are also in an optimum state of adaptability. Ks between eight and 10, Kauffman believes, correspond to the 'edge of chaos' – the area where complexity rules. For Kauffman, natural selection merely takes organisms and, indeed, the whole of life on earth, toward an ideal state:

Much of the order we see in organisms may be the direct result not of natural selection but of the natural order selection was privileged to act on. Second, selection *achieves* complex systems capable of adaptation... There are general principles characterising complex systems able to adapt: They achieve a 'poised' state near the boundary between order and chaos, a state which optimises the complexity of tasks the system can perform and simultaneously optimises evolvability.[43]

Simultaneously, the organisms that exist, their form, correspond to universal types that are in essence pre-given. For Kauffman, conventional thinking on ontogeny only supplies the *how* of pattern formation, not the *why*.

Writing of his overall view of 'the manifold marriage of self-organisation and selection', Kauffman describes the marriage not just as a result of selection 'achieving systems best able to adapt', but also of the 'emergence of the generic properties of that ensemble as quasi-universals in biology'.[44]

Kauffman: Practical Doubts

Kauffman's work is the culmination of 30 years writing, thinking, and arguing. It is a highly readable, audacious challenge to orthodoxy. His treatment of the edge of chaos in the 'NK' model is the most precise, and credible, application of this central aspect of complexity theory. From a mathematical point of view, it makes a lot of sense. But is it biology? We begin by looking at some empirical objections to his ideas.

The notion that a value for K of between eight and 10 is the value to which evolution would always be likely to take natural systems, because these values put an eco-system on the edge of chaos, and thus optimise adaptability, is perhaps the central claim of Kauffman's *Origins*. And yet, it is a claim with very little empirical support. Here, after an interview with Kauffman, is how Waldrop describes it:

The evidence consists of a sort of power law in the fossil record suggesting that the global biosphere is near the edge of chaos; a couple of computer models showing that systems can adapt their way to the edge of chaos through natural selection; and now one computer model showing that ecosystems may be able to get to the edge of chaos through coevolution. 'So far', he [Kauffman] says, 'that's the only evidence that I know that the edge of chaos is actually where complex systems go in order to solve a complex task. It's pretty sketchy. So, while I'm absolutely in love with this hypothesis – I think it's absolutely plausible and credible and intriguing – I don't know if it's generally true'.[45]

Evidence for biological complexity is, then, 'pretty sketchy'. In *The Origins of Order*, Kauffman again repeated that caution was necessary:

> The range of coupled landscapes that exhibit a phase transition from an ordered to a chaotic regime is unknown. The generality of the claim that the phase transition affords the highest fitness is unknown. The efficacy of selection to achieve and sustain the edge of chaos against drift is unestablished. The applicability to real coevolving systems is untested.[46]

As if all these gaps in knowledge were not enough, Kauffman's application of the principles of complexity to the problems of ontogeny is similarly lacking in empirical foundation.

As a practical example of his theory, Kauffman puts forward a new account of pattern formation in the developing fruitfly *Drosophila Melanogaster*. And yet he concedes: '*no single adequate theory* of pattern formation in *Drosophila* is in hand'.[47] And if the patterns that arise in the development of Drosophila are badly understood, we can assume that little else in ontogeny is well understood, since Drosophila is the most studied of all organisms.

It is Kauffman's belief that 'even the outline' of an adequate theory of ontogeny is 'not available'.[48] Evidence is missing;

theory, even with the humble fruitfly, is weak. Yet despite all this, Kauffman is forthright in claiming a universal application for his own theory.

Kauffman's argument is that life is an inevitable consequence of principles of complexity operating on and through non-living matter. He believes that the forms of order achieved through evolution, and the patterns developed through ontogeny, are essentially independent of the specific laws of physics, chemistry and biology which govern the component parts of organisms. This is a good example of the application of the idea of 'emergence'. However, Kauffman does not in fact show that the specific laws can be so disregarded. This is not surprising, since he has done no more than postulate a mathematical model, albeit an appealing one.

Kauffman could, of course, respond that the problems he is grappling with are ones that as yet have no adequate explanation. And he would be right in some areas. The problem with his approach is that it makes a *virtue* out of ignorance. He has not shown that the underlying laws can be ignored; yet he claims that the advantage of his system is that we can save ourselves the effort of grappling with the underlying laws and their effects:

From a practical point of view, we may never be able to work out the details of the genomic regulatory system in a single inbred higher eukaryote. Even were we to succeed, the fluidity of the genome means that the regulatory system in neighbouring organisms, sibling species, and so forth, are dancing away from us faster than we may ever be able to grasp them. The practical epistemological problem, in short, is that we may never be able to carry out the reductionist dream of complete analysis but will want nevertheless to understand how these systems work. To the extent that known local properties engender an ensemble whose typical members exhibit many of the properties found in differentiation and ontogeny, then to that extent those properties do not depend upon the details of the genomic system. Therefore we can legitimately *explain* those properties by understanding why they are typical of the ensemble in question. Indeed, the typical properties of such ensembles emerge as the new 'macroscopic' observable of this new kind of statistical mechanics.[49]

For Kauffman, an organism's properties are to be explained not by genomic details, but through the statistical emergence of properties from an ensemble of ensembles.

However, Kauffman has still to demonstrate at all convincingly that the properties do not depend on the underlying laws of physics, chemistry and biology. So his fondness for properties which are 'typical of the ensemble' fails to amount to an explanation of anything. Kauffman has merely made a virtue out of *not* carrying out the 'reductionist dream'.

In Praise of 'Junk DNA'

The way Kauffman abstracts from the laws of physics, chemistry and biology, without actually showing that it is legitimate to do so, is a substantial defect in his approach. There are also defects of substance beyond the simple failure to produce empirical evidence.

First, the 'NK' model, and the idea of fitness landscapes, make an illegitimate abstraction from the possibility of N increasing. Illegitimate in the sense that N increasing is central to the way organisms develop. Recall that Kauffman argues that 'the fitness of the entire genotype is the mean of the fitness contribution of its N sites'.[50] This amounts to averaging over N. Only in this way does the size of N itself not matter, and only because of this does a given value of K become the edge of chaos.

With N effectively fixed, Kauffman argues that a given value of K will be optimal because a lower value will create a smooth landscape upon which natural selection cannot hold organisms at the peaks of fitness. A higher value of K, on the other hand, leads to a very rugged fitness landscape – one upon which peaks of fitness fall toward the floor of the valley of the landscape, and hence one in which the highest attainable fitness is too low.

The argument is persuasive, but it is only really persuasive if we accept the way Kauffman abstracts from N increasing. For if N is allowed to increase, a different result follows.

For example, if K increases *as a necessary consequence* of N increasing, the higher N may well compensate for the more rugged landscape. How? By raising the fitness level of the valley floor above the old fitness level of the peaks. There are endless other possibilities, once the unlikely assumption of N being fixed is dropped.[51]

It is not necessary to imagine that an organism magically enlarges its genome to create a greater N. Much of the DNA of an organism is thought to be 'Junk DNA': that is, DNA which plays no functional role. As an organism changes, however, it can draw upon this DNA and turn it into functioning DNA, thus enlarging N.

This possibility better fits the real patterns of evolution. It is also truer to evolution as envisaged by Darwin, who considered organisms to be not so much ideal types, as contraptions that were cobbled together. In a review of Kauffman's book in *Nature*, Gabriel Dover ironically remarked that designing life through ensemble theory was 'as cosy as designing a jumbo jet on a computerised drawing board'. Yet the real trick of life, Dover went on, was 'the gradual evolution of a jumbo from the first wire and wood contraptions while the bloody things are still flying in the air!'. To explain this 'trick', Dover pointed to 'the internal tolerance governing evolving molecular interactions', the basis of which are 'modular and redundant systems'.[52]

The point about modularity and redundancy in genetics, about 'Junk DNA', is that they allow evolution to make new forms. It follows from this that while there are forms and patterns in evolution, they are *not pre-given*, as Kauffman argues. Rather, natural selection creates form. It does much more than actualise existing possibilities. As the evolutionary biologist Stephen Jay Gould puts it:

Any replay of [evolution's] tape would lead evolution down

a pathway radically different from the road actually taken. But the consequent differences in outcome do not imply that evolution is senseless, and without meaningful pattern; the divergent route of the replay would be just as interpretable, just as explainable *after* the fact, as the actual road... Each step proceeds for cause, but no finale can be specified at the start, and none would ever occur a second time in the same way, because any pathway proceeds through thousands of improbable stages.[53]

There *is* order in Gould's portrait; but it is an order which, because of its reliance on accidents, is a post-hoc affair.

Coevolution and Gaia

The second, and perhaps more serious, defect with Kauffman's 'NK model' is that he abstracts from natural selection by ignoring coevolution. The result is that, when he does turn to the subject of coevolution, he transforms it into Gaia.

How does Kauffman do this? The 'NK' model is built by presuming a relatively fixed natural environment. This is a reasonable assumption as far as the *physical* environment is concerned: after all, the physical environment changes at a much slower rate, in general, than organisms do. But what the assumption ignores is that, for each organism, the surrounding environment also includes a whole range of other organisms which are not fixed at all.

The theory of coevolution takes account of this. It recognises that all the organisms on earth, as well as the physical environment, evolve together, while maintaining their independence. By taking both the physical and the biological environment to be relatively fixed, the NK model thus abstracts from coevolution in its very construction. It therefore necessarily fails to capture the essence of the evolutionary process, since coevolution *is* evolution. Kauffman then applies his model to coevolution. Inevitably, he gives a false representation of coevolution, since the model was constructed by ignoring it.

Kauffman's analysis of coevolution proceeds like this. First, he applies the 'NK' model to the whole ecosystem. That system, Kauffman finds, is perched, Gaia-like, at 'the edge of chaos'. But this application to the whole is itself an anti-Darwinian, Gaia-inspired, turn; the whole system behaves like Gaia because Kauffman's 'edge of chaos' study of it is just Gaia by another name.

Second, Kauffman tries to bring coevolution in by arguing that this edge of chaos is reached by one organism 'tuning' another as they move, together, toward it. Kauffman seems blind to the fact that what is optimal for one organism may well be sub-optimal for another. In nature, organisms do closely interact, but this is usually by one being parasitic upon the other. Kauffman's idea of 'tuning' also directly contradicts Darwin's insistence that no organism can evolve for the good of another.

Kauffman's abstraction from coevolution leads to a misleading picture of the evolutionary process. His picture shares some of the defects of that school which favours the existence of what it calls *Artificial Life*. The idea here is that computer programmes, or little 'animals' on a computer screen, are a form of life if they can interact with each other and 'evolve'. Not for the first time, here, we find a mix-up of metaphors going on. What the A-Life school ignores is that the physical environment for these new 'life' forms is fixed (it is created by humans) and does not evolve with the 'organisms'.

In one sense, though, the A-Life theorists have made a better model of evolution than has Kauffman: at least their little beasts coevolve with each other. In Kauffman's world, by contrast, each species only interacts with others once its development path is set. Yet that setting is performed by first isolating the organism from any genuine coevolution.

The various defects in Kauffman's application of complexity theory to biology are replicated in other attempts to apply the theory to real-life situations. In the section which follows, we concentrate on the lack of empirical proof, not only in genetics, but also in economics. In a further section, we then present some theoretical objections to the idea that complexity theory represents a new universal theory of order in nature.

Poor Evidence for Complexity in Genetics and Economics

Complexity theory, like chaos theory, is fascinating mathematics. But we strongly doubt that it models more than a few physical systems. Little evidence has been produced to show that complexity can explain physical mechanisms in natural systems. Complexity theorists often claim to provide explanation of physical mechanisms. But typically they have done nothing more than re-state a fact about order without explaining *how* there is order.

Consider, for example, the theorists' views on the nature of DNA. During the Second World War, Erwin Schrödinger turned his mind to biology and produced the ground-breaking pamphlet *What is Life?* Using the principles of quantum theory, he postulated that the building block of life would have to be a stable chemical molecule which gave life the ability to draw a stream of order from the environment. His work proved prophetic. It inspired a generation of biologists, including James Watson – the man who co-discovered Schrödinger's molecule in the shape of DNA.

Complexity theorists claim that their work – studying the emergence of order from chaos – allows us to see the nature of DNA in a broader picture, as one example of self-organising, or 'dissipative', systems. But it does not. All they show is what Schrödinger already knew to be a brute fact – that DNA has the ability to create order locally, *without contradicting the second law of thermodynamics.*

What about complexity in economics? As with the attempt to place DNA in a 'broader' context, and as with the use of chaos theory in the study of economic trends, no mechanisms have been explained by complexity, and no phenomena associated with complexity have been shown to exist.

At the Santa Fe Institute, Gell-Mann believes that the modelling has been a great achievement. But he admits that it has yet to come up with any hard proof:

> The economics programme has been one of the most successful activities of the Santa Fe Institute, in terms of stimulating new theoretical and modelling activities of high quality. Ultimately, of course, success must be measured, as in all theoretical science, by explanations of existing data and by correct predictions of the results of future observations. The Institute is still too young and the problems it studies too difficult for very much success of that kind to have been achieved so far.[54]

Gell-Mann is too kind. Young as it is, the Institute has been in existence for more than 10 years, and many of the people at it have been trying for even longer to use complexity to model economics. In any case, that problems are very difficult does not excuse a failure to make empirical sense of 'self-organisation', 'the edge of chaos', and other categories associated with complexity theory.

The British economist Paul Ormerod has made an attempt to apply complexity theory in his book *The Death of Economics*. Ormerod wittily castigates modern economics. He attacks the discipline for its reliance on linear models, for its blindness to real data, and for its disdain for empirical proof. His alternative? Non-linear mathematics in general, and complexity theory in particular.

However, Ormerod does not field any answers. To very fundamental questions, like the roots of mass unemployment today, he has no replies. All he does is postulate a non-linear relationship between variables such as unemployment and rates of profit. Even if his mathematics is correct, he has offered no *explanation* of these relations.

More generally, just as Kauffman gives no explanation of why it should be possible to ignore the laws of physics, chemistry and biology in constructing a theory of order in biology, so complexity theorists give no convincing explanation of what 'emergence' might mean in nature as a whole. Take, for example, the treatment of emergence by Cohen and Stewart in their book *The Collapse of Chaos*.

Cohen and Stewart believe that the laws of sub-atomic physics have little or no relevance to the workings of biology, or the universe. They may be right; but they do not *show* why they are right. Reductionism is once again skewered, and 'emergence' does indeed take place. But at no time do Cohen and Stewart state how emergence does or does not give rise to other laws in nature.

Perhaps we find order in nature fascinating because we would not exist without it. Whatever the reason, fascinating it is. But from Prigogine through Kauffman to Gell-Mann, *explanations* are lacking of *mechanisms*. Complexity theory suggests that order is intrinsic to natural law. But the mechanisms which might prove this appear to be missing. At best, the various authors describe how ordered systems function.

Will complexity theory ever come up with a convincing account of a real system in the future? Apart from our practical objections, there are theoretical reasons to doubt that it will.

Complexity: Theoretical Doubts

The existence of complexity may in many cases be something which does not require explanation. Reflecting on the lessons of evolution on earth, Gould points out that there is no reason to assume that any 'principle of complexity' is at work in nature. Such thinking is a result of anthropic reasoning: 'we look at complexity the wrong way because we are complex and we tend to see a drive in that direction.'[55] Gould goes on to point out that the existence of complexity may simply be a consequence of simple beginnings:

> While not denying that the most complex creature on earth has tended to be more complex through time, we are really looking at it in an odd way... it is more an expansion from the necessary simplicity of beginnings than a necessary predictable drive towards higher states of organisation.[56]

Just in case we forget, Gould reminds us that the most common organisms on earth are bacteria: organisms which have been around in a similar form since not long after life on earth began. The prevalence of bacteria alone suggests that if there has been a drive, in evolution, toward greater complexity, it is a drive which has passed a significant part of the ecosystem by.

Gould's is only a suggestion. But his hypothesis fits well with Darwinian theory, for which there is a working model of physical and biological mechanisms. To postulate a wider 'principle of complexity' is merely to restate the fact that the most complex organisms on earth have become more so with the passing of time.

Of course, order does exist in nature. While a general theory of order may not be needed, order is still a problem worthy of investigation. In this light, there is a further and more central theoretical doubt that can be raised to complexity theory as a theory of nature. Many of the things which it claims are intrinsic to natural law may well be to do with the interaction of three things: discrete natural laws; chance happenings, and the effects of boundaries.

The patterns achieved by evolution on earth are more convincingly explained by the subtleties of coevolution, involving these three factors, than they are by abstract theories of pattern formation.

As a theory, coevolution is essentially dependent on the existence of a boundary. The Earth, with a continous input of energy from the sun, provides a boundary within which evolution occurs.

Generalising from this example, coevolution might provide a good example of the working of natural law in general. Just as evolution works through the interaction of organisms and their physical environments, so order might well be the result of distinct natural laws and the effects of boundaries. Alternatively, order might be the joint result of a universe which was explicable by a Theory of Everything, together with frozen accidents. These are just suggestions. However, both frameworks have generated more insight into nature than have any 'principles of complexity'.

Why Complexity Cannot Capture the Uniqueness of Social Systems

If order is context-specific, then a general theory of order is not possible. To try to create one, and use it as a guide to studying different systems, is to risk failing to explain anything about specific systems; and this is especially the case when we move from the study of nature to the study of human behaviour and human society.

Gell-Mann thinks that complexity theory can encompass, and explain, human learning and development. In the future, he says, once the transition to a 'sustainable society' has been achieved, nature and society will merge together to form one whole complex adaptive system, subject to a common order. Prigogine also believes that the same laws of self-organisation can explain both nature and society. In his *Order out of Chaos*, jointly written with Isabelle Stengers, he says:

> Present-day research leads us further away from the opposition between man and the natural world. It will be one of the purposes of this book to show, instead of rupture and opposition, the growing coherence of our knowledge of man and nature.[57]

In our view, nothing could be further from the truth. The more science pushes ahead, the more we realise the uniqueness of humanity. Lamenting 'rupture' and 'opposition', Gell-Mann and Prigogine want to lower *humanity* to the level of *nature*. In the process, they efface the specifics of both.

Humanity has made society, but it has not made nature. Human action is fundamentally different from any law of nature in that it is animated by *purpose*, or *will*. Human society has a flexibility that no natural system has, because humans can *plan ahead*. Further: human learning is, as a consequence of this, totally different from animal learning. All these are crucial differences between humanity and nature.

Animals have only a limited ability to learn from experience. In his book *The Making of Memory* (1992), Steven Rose, a British geneticist and a long-time critic of sociobiology, gives an example of how animal learning places limits on animal behaviour.

In his laboratory, Rose found that he could train a chick to avoid pecking a coloured bead by coating it in a bitter-tasting substance. He then tried a different approach, and obtained a different result:

> I tried to pair the bead-pecking with a different form of discomfort. I arranged the experiment so that every time the chick pecked a dry, tasteless bead, it felt a mild electric shock to its feet.[58]

This new arrangement did not stop the chicks pecking the bead. If anything, they did it more. In other words, the whole experience was one they were unable to learn from.

What is true of chicks is true of all animals to a greater or lesser extent. Everything depends on context, and the complexity of the animal. However, there is a fundamental passivity in animals' relationship to their own experience which means that their behavioural patterns change very slowly, if at all.

Rose reasons that the inability of his chicks to learn from their new experience was due to the fact that 'in nature they are scarcely likely to have to learn the peck-pain-in-the-foot relationship'.[59] Through evolution, they have not developed the behavioural flexibility to deal with it. In other words, the evolved genetic make-up of the chicks has not even given them the behavioural flexibility to respond in a more 'intelligent' manner to an electric shock. With animals, then, genetically-encoded behaviour may not be a rigid prison; but it is certainly limited in relation to 'learning new tricks'.

Humans are not so constrained. Humans theorise and adopt new behavioural responses accordingly. After 10 pints of beer, even otherwise rational people might keep on pecking like the chick; but if they did not have a theory of what was giving them electric shocks to start with, they would when sober once more. And no human would begin pecking again until he or she worked such a theory out.

Unlike even the most flexible of animals, the apes, human beings can absorb lessons from experiences and experiment with new patterns of behaviour. Humans theorise, and they act purposively. They can do this because they have acquired purposive behaviour through growing cultural and social organisation. Humans, and only humans, can draw upon the accumulated, collective knowledge of international society, past and present. Animals do not pass on their experiences from one generation to the next; certainly not in a cumulative way. Humans do. That unique circumstance has created a unique learning pattern.

Social organisation has given humans the capacity to move beyond natural evolution and to overcome the limitations of their biological make-up. In the words of Richard Lewontin, 'social organisation does not reflect the limitations of individual biological beings', but is rather 'their *negation*.'[60] In Kauffman's language of fitness landscapes, we might say that, unlike animals, humans beings are *not* condemned to drift on optimal terrain at the edge of chaos, with the optimum conditioned by the environment in which they live. Humans can and do *re-design* the environment and thus change the landscape. They can also change the landscape by changing the way in which they relate to each other.

The history of human civilization has been the replacement of one kind of society by another. These changes are not akin to complex physical systems moving from one state to another, nor are they analogous to phase transitions in nature. They are examples of human progress. The dynamics of the social are quite different from those of the natural.

In the future, human society could never form one whole complex adaptive system with nature, as Gell-Mann believes. Of course, the relationship between humanity and nature will

change; but humanity will always have a fundamentally instrumental relationship to nature, because humanity has
purposiveness, and nature does not.

Humanity needs nature. Our stress on the uniqueness of
human beings makes us, for example, keen to preserve the
widest variety of species in nature. But humanity does not just
need nature; it will also, for some time to come, be profoundly
affected by dramatic changes – by earthquakes, tornados and,
perhaps, by global warming. Purposiveness, however, gives
humans the capacity for using *innovation*, as well as conservation, to deal with change. Men and women are not passive
victims of natural change, whose only hope is that conservation
will minimise disruption. Humans aspire to take control of their
surroundings and mould them to their advantage.

The Scientific Appeal of Chaos and Complexity

How can we account for the popularity of both chaos and
complexity among scientists? The appeal of chaos and
complexity theories to mathematicians is obvious. All mathematicians, a cynic might argue, are Platonists to one degree or
another. Chaos and complexity are mathematically very elegant,
and this is appealing. Of chaos, Ruelle writes:

> I have not spoken of the aesthetic appeal of strange
> attractors. These systems of curves, these clouds of points
> suggest sometimes fireworks or galaxies, sometimes strange
> and disquieting vegetal proliferations. A realm lies there of
> forms to explore, and harmonies to discover.[61]

And again: 'One can imagine the system endlessly tracing out
patterns within patterns within patterns'.[62]

For the mathematician as aesthete, this is fair enough. But
what then is the appeal of chaos and complexity to physicists,
and to other natural scientists? First and foremost, scientists
tend to turn to these particular non-linear models out of
frustration with other approaches. It is this, rather than any

great explanatory power, which explains the 'draw' of chaos and complexity theories.

Take the example of fluid mechanics, one of the most frustrating areas of applied mathematics, and an area in which much effort is being applied to find chaotic patterns. In fact turbulence in fluids has bedeviled generations of scientists. In 1932, the British physicist Horace Lamb told a meeting of the British Association for the Advancement of Science:

> I am an old man now, and when I die and go to heaven there are two matters on which I hope for enlightenment. One is quantum electrodynamics, and the other is the turbulent motion of fluids. And about the former I am really optimistic.[63]

David Ruelle developed a chaos-based model of turbulence in the early 1970s. In doing so, he has a claim to inventing the term 'strange attractor'. But, as he himself notes, all he has explained is the *start* of turbulence, and not turbulence itself: 'Strange attractors and chaos have clarified the problem of the onset of turbulence, but not that of fully developed turbulence'.[64]

Scientists have turned to chaos theory for answers. But the key problems in fluid mechanics have proved as difficult to solve as ever. Chaos theory has offered no big, new insights. Ian Stewart, another enthusiast, gives this account of the problems:

> Much of turbulence remains a mystery. Fully developed turbulence, if it involves strange attractors at all, may require attractors of enormous dimensions – a thousand, a million. At the moment we can say nothing worth knowing about these. Many turbulent effects seem to be connected to boundaries – the walls of pipes, for example – and strange attractor theories haven't yet been related to the influence of boundaries.[65]

In fact, the motivation to use chaos theory to model fluids turns out to be based not so much on evidence, as on exasperation with other methods and approaches.

Of course, for scientists, feelings of frustration with existing methods are unavoidable. But there is a danger in letting impatience come to distort assessments of alternative approaches to the mathematical modelling of physical systems. This procedure is in fact quite widespread in the literature on the applications of chaos. A chaotic approach is justified in terms of the failure of *other* approaches. We have a *negative proof-procedure*.

As a further example of the process-of-elimination logic behind the growing use of chaos and complexity, take the attempt to apply non-linear mathematics to economics. In economic theory, it is the fashion now to attack the idea of equilibrium. Given the turbulence in stock markets and economies since 1987, this is hardly surprising. However, in their pamphlet for the Institute of Economic Affairs, David Parker and Ralph Stacey go further. They commend chaos theory as a model of economic systems simply because it does not predict equilibrium:

> We reach, then, one of the main insights chaos theory brings to organisations. In most of the management literature, the concept of 'managing change' is built on the idea that stability is desirable... The insights of chaos theory show just how limited this idea is. If we want organisations to be innovative, we have to accept that change is continuous.[66]

In fact, we do not need chaos theory to realise that change in organisations, and specifically in the capitalist firm (which is what Parker and Stacey are interested in), is continuous. Neither does the currently palpable but in fact longstanding failure of equilibrium economics prove that economics can be modelled by chaos theory, which is the implication Parker and Stacey wish to draw.

Gell-Mann, at least, advocates 'making economics less dismal'. But in this task he wants to use complexity theory, on the grounds that

> a preoccupation with a kind of ideal equilibrium, based on perfect markets, perfect information, and perfect rationality of agents, has characterised a great deal of economic theory during the past few decades [without explaining how the real world works].[67]

Gell-Mann is right about the bankruptcy of neo-classical economics. But this is not an argument for a complexity-based theory of economics, as Gell-Mann seeks to imply.

Complexity as a New Religion, and as a Rediscovery of Conservatism

In concluding our treatment of complexity here, we note that a focus on the growth of complexity in nature, as an *inherent property* of natural *process*, has strongly religious and backward-looking antecedents. In this sense, complexity theory, as a theory of nature, is nothing new, and has an obscurantist character which true science would do well to be wary of.

In the 1920s, the British mathematician and philosopher Alfred North Whitehead developed what he called 'process' philosophy. Nature was 'a structure of evolving processes'. For Whitehead, the reality was the process.[68]

As a result of this, nature had the character of *becoming* as well as *being*. Whitehead argued that an individual element could express the unity of the process determining the development of the world:

> Every actual occasion exhibits itself as a process. It is a becomingness... It also defines itself as a particular individual achievement, focusing in its limited way on unbounded realm of eternal objects.[69]

Finally, unity as process was, for Whitehead, God: 'God is not concrete, but He is the ground for concrete actuality. No reason can be given for the nature of God, because that nature is the ground of rationality'.[70]

From Whitehead's 'becomingness' it is but a short step to complexity's alleged phenomenon of 'emergence'. The decline of *organised* religion in Western society should not prevent us from recognising the fact that many of today's *inclinations* are of a religious type. In this respect, belief in the universal applicability of complexity theory may be more a social symptom than an illuminating analysis.

Another example, apart from Whitehead, confirms the theistic heritage behind complexity. The views of Teilhard de Chardin are a part of the Judeo-Christian tradition which gives to humanity a privileged position in the universe. Like Whitehead, de Chardin incorporated 'process philosophy' into his perspective. In his famous *The Phenomenon of Man* (1955), he argued that humanity was the pinnacle of natural self-organisation: humankind was the 'last and highest form of aggregation, in which the self-organising effort of matter culminates in society as capable of reflection'.[71]

If humanity was the pinnacle of self-organisation for de Chardin, God was the process itself. As a consequence,

> evolution has come to infuse new blood, so to speak, into the perspectives and aspirations of Christianity. In return, is not the Christian faith destined, is it not preparing, to save and even take the place of evolution?[72]

The idea that nature has the character of *becoming* as well as *being* is central to Prigogine's work. It is not far from here, then, for advocates of complexity to follow de Chardin and see evolution as the unfolding of God's work. Supporters of Prigogine's views, indeed, do exactly this. As Erich Jantsch put it: 'God is thus not absolute, he evolves himself – he is evolution.'[73]

While 'self-organisation', 'the edge of chaos' and complexity theory in general represent perfectly decent mathematics, the impulse to claim that they represent a unified theory of humanity and the natural world is undoubtedly closer to religion than to science. As we noted at the end of Chapter Two, Brian Arthur claims that complexity theory has much in common with Eastern philosophy. Could it not be, rather, that he is attracted to Eastern philosophy, and that this in turn leads him to claim that complexity theory is a new universal theory of nature?

Our alternative view of the universe and of process is very different from that presented by enthusiasts for complexity theory. It incorporates the Darwinian view of *natural* evolution, while proposing a different theory of *human* development.

Darwinism tells us that in terms of *natural evolution*, human existence is but a matter of chance. There is no cosmic plan; humanity represents a contingent event. Here, for our part, we agree. We accept the contingency of natural evolution.

But if we accept the contingency of humankind *in an aeons-long evolutionary context*, we do not accept that *human nature now* is simply a biological fact. Rather, as humanists, we stress the way humanity has made and continues to remake itself. We observe how rational enquiry, intervention in the real world, and history have all, in a very short time, by evolutionary standards, completely transformed people. In this framework, humanity is unique compared with both animate and inanimate matter.

Our alternative, irreligious view of the universe does not just extend to evolution, but to economies too. Here again we find that the apostles of contemporary complexity have allies who do not inspire confidence.

In his last major work, *The Fatal Conceit: The Errors of Socialism* (1990), the Austrian economist Friedrich Hayek turned to natural science. Hayek had long thought that individual, capricious responses to market forces made humanity largely impotent before their ceaseless interplay. In *Fatal Conceit*, he argued that Reason was further constrained by the complexity of nature.

In his valedictory tract, Hayek registered, with not a little satisfaction, the growth of a new attitude to the working of natural law:

When I began my work I felt that I was nearly alone in working on the evolutionary formations of... complex self-maintaining orders. Meanwhile, researchers on this kind of problem – under various names, such as autopoiesis, cybernetics, homeostasis, spontaneous order, self-organisation, synergetics, systems theory, and so on – have become so numerous that I have been able to study closely no more than a few of them.[74]

For Hayek, natural phenomena which were founded, at bottom, by *feedback*, and which were also described by a number of *opaque nouns*, were wonderful allies.

Hayek had a keen sense for those scientific theories which could be of use to him. He wanted to set boundaries to human potential in economics; now he conscripted complexity theory and self organisation into his campaign.

That complexity theorists should be supported by an ideologue like Hayek does not, of course, prove that they are wrong. But insofar as they share his basic imperative to humanity – Submit to Forces Beyond Your Control – the similarities are clear enough. The 'discovery' of complexity in the world appears to have a lot going for it. But it can all too easily prove to be a rediscovery of conservatism.

Chapter Five:
Science and Humanism

From the sixteenth century onwards, a series of breakthroughs in cosmology, astronomy and other fields had an enormous impact on human thinking and practice. The impact of the scientific revolution was indeed so great that the English historian Herbert Butterfield portrayed it as 'a civilisation exhilaratingly new perhaps, but strange as Nineveh and Babylon'. Apart from the rise of Christianity, Butterfield held, there was 'no landmark in history' worthy of comparison with the Scientific Revolution.[1]

The Scientific Revolution represented the triumph of rationality and experiment over the metaphysics, superstition and speculation that had gone before. It was more than simply an advance in scientific knowledge: it was a part of a wider shift in attitudes and beliefs. The Scientific Revolution was the product of dynamic social progress, and at the same time an essential contributor to that progress.

From the seventeenth to the nineteenth century, it was a common assumption of the intelligentsia that scientific advance and social progress marched hand in hand. This consensus reflected the sense of confidence which prevailed in a period of unprecedented economic and social development.

Since it took place, there really has been no landmark in history to rank with the Scientific Revolution. The twentieth century has, after all, witnessed regular economic stagnation and war. Today's intellectual climate is one that is deeply perturbed – about the future of both science and society. In particular, the view that scientific advance leads to social progress has been widely questioned, and occasionally repudiated altogether. There has also been a reappraisal of the Scientific Revolution, and, in some cases, an outright attack on scientific rationality. Chapter Six looks at developments in the twentieth century. This chapter examines the inter-relationship between scientific advance and social progress in the three preceding centuries.

Isaac Newton: 'Nearer the Gods no mortal may approach'

It is impossible to date exactly either the beginning or the end of the Scientific Revolution. However, most commentators accept that the most significant period lies between the death of Nicolaus Copernicus in 1543, coincident with the publication of his *De Revolutionibus Orbium Caelestium* [On the Revolution of the Heavenly Spheres], and the publication of Newton's *Principia Mathematica* in 1687, which itself was assisted by the work of Galileo Galilei [1564–1642] and Johannes Kepler [1571–1630].

Butterfield's argument is that the Scientific Revolution was the catalyst for the wider social changes that occurred at the time. His thesis is seductive, because the influence of scientific ideas and practices on the key thinkers of the period is undeniable. Nobody influenced the thinkers of the Enlightenment more than Newton. As the poet Alexander Pope wrote, in 1735, in his famous epitaph for the man:

Nature and nature's law lay hid in night:
God said, Let Newton be! and all was light.

These two lines alone bear testimony to the awe in which Newton was held in his own age.

Born in Woolsthorpe, Lincolnshire, on Christmas Day 1642, Isaac Newton became the pre-eminent figure of the scientific revolution. After graduating from Cambridge in 1664, he spent the next two years back home, escaping the Great Plague. It was during this enforced retreat that, according to Newton, he did his main work – the culmination of which would make him the most renowned and influential scientist until Albert Einstein.[2]

By 1669, the then Lucasian Professor of Mathematics at Cambridge, Isaac Barrow, had been so impressed by Newton's work that he resigned his chair in Newton's favour. Newton's most famous and important work, *Philosophiae Naturalis Principia Mathematica*, was not published until 1687.[3] In it he demonstrated how, from a few basic principles, the behaviour of physical objects – terrestrial and celestial – could be predicted and comprehended. His contemporary, Edmund Halley [1656–1742], believed of Newton's *Principia* that 'Nearer the Gods no mortal may approach'.[4]

Newton's influence quickly spread to all areas of scientific thinking. Jean Lerond d'Alembert [1717–83], the French mathematician, exuded confidence in science when he wrote:

> Our century is called.... the century of philosophy *par excellence*... The discovery and application of a new method of philosophising, the kind of enthusiasm which accompanies discoveries, a certain exaltation of these ideas which the spectacle of the universe produces in us – all these causes have brought about a lively fermentation of minds, spreading through nature in all directions like a river which has burst its dams.[5]

Interest in Newton spread not only to continental Europe and beyond, but also beyond scientists. The scale of popular interest was immense, and rather puts in perspective the best-selling status of Stephen Hawking. The eighteenth century saw many writers producing accounts of Newton's work. Titles included *Astronomy Explained upon Sir Isaac Newton's Principles, and Made Easy to Those who have not Studied Mathematics*. There were also political tracts, such as *The Newtonian System of the World: the Best Model of Government*. Finally there were works with strange titles, such as *Newtonianism for Ladies*.

Bernard Le Bovier de Fontenelle [1657–1757], Secretary of the Paris Academy of Sciences, believed that the new 'geometrical spirit' of Newtonian physics could improve fields of human endeavour as varied as politics, morals and literary criticism. This opinion was shared by the thinkers of the Enlightenment and by those who made the revolutions and framed the new constitutions of the day.

Both of the two major political revolutions of the eighteenth century, whose reverberations are still with us, had a definite relationship with the Scientific Revolution. The American revolution culminated in the Declaration of Independence in 1776; the French revolution started in 1789 with the fall of the Bastille and the Declaration of the Rights of Man and of Citizens.

Both drew sustenance from the thinkers of the French Enlightenment, who themselves were greatly impressed and influenced by the development of science in the seventeenth century and the work of Newton in particular.[6] Thus the British philosopher John Locke [1632–1704], who influenced the American revolution, regarded himself as merely an 'under-labourer' to the 'incomparable Mr Newton'. It is easy to see the appeal of Butterfield's idea that the Scientific Revolution *caused* the subsequent political and economic transformations.

How Society Shaped Science

But what are we to make of Butterfield's thesis? There can be no doubt that science *has* had a dramatic impact, directly and indirectly, on the nature of society in all ages. It has strengthened certain ideas: for example, the idea of progress and the feeling that people can control both nature and their own destiny. Science has also changed forever humanity's perception of its place in the cosmos. From being the inhabitants of a body at the centre of the universe, around which the rest of the universe revolved, we now see ourselves as the habitants of a tiny planet. We revolve around an ordinary star, on the fringes of a galaxy which is one among countless others.

The impact of science on our perceptions of the world and ourselves was never more pronounced than during the period of the Scientific Revolution and the Enlightenment. In many ways, humanity continues, even in the late 1990s, to deal with the effects of this period.

Clearly, there exists a very important interplay between the development of science, its impact on society, and its impact on society's visions of the future. Yet which should be regarded as Prime Mover? We believe that Butterfield was wrong to elevate science above social change in the way he did.

The growth of trade and society predated the development of progressive ideas and the Scientific Revolution. Moreover, the broad sweep of human history shows that the centres of techno-logical and scientific excellence in the world have shifted to *follow* the centres of commerce and industry.

From 400 AD to 1000 AD Europe was, in terms of scientific achievement, a backwater.[7] Happily, some of the high-points of Greek science were kept alive in the Arab world, and some devel-opment of mathematics beyond that known to the Greeks took place there.[8] From then on it was the growth of Italy's trade with the Arab world in the eleventh century which created the auspices for the reintroduction of Greek science into Europe. The primacy of social factors is in many ways clearest in this early period. Without the development of trade and the transformation in society it brought with it, there would have been no basis for Europe developing as the centre of science in the way that it did.

After a long and faltering start, the Italian Renaissance, by the time of the fifteenth century, began to flourish.[9] In technical subjects, art, and the humanities, Italy led the world. However, the Renaissance ended quickly, after the sack of Rome in 1527 by Spain and France; and its peak, by then, had long since passed. The discovery of America in 1492 and the shifting of trade routes to the Atlantic seaboard dislocated Italy from her previously central position. While Galileo and other Italians continued to press their country's scientific contributions upon the world, England, Holland and France began to become centres of scientific excellence in their own right. Certainly, scientists in these countries enjoyed greater freedom to carry out their work.[10]

A striking feature of the Renaissance and after was the merging of intellectual activity with its practical applications. Writing in the middle of the sixteenth century, J. Fernal drew attention to both the technological achievements of the period, and what he termed 'the restoration of scholarship':

> The world sailed round, the largest of Earth's continents discovered, the compass invented, the printing-press sowing knowledge, gun-powder revolutionising the art of war, ancient manuscripts rescued and the restoration of scholarship, all witness to the triumph of our New Age.[11]

The unique blend of theory and practice which was distinctive of science in the New Age followed from the transformations going on in the social context surrounding that science. Humanity had fresh, ambitious and practical goals to achieve – not least, in international navigation.

The differences between post-Renaissance and ancient Greek science also bring out the decisive role played by social advance in shaping the character of scientific ideas. Here, Morris Kline, a historian of mathematics, has done much to explain the ancient Greeks' unrelenting emphasis on deductive logic and abstract thinking. For Kline, that emphasis was a consequence of a static society built on slavery: practical work was totally alien to ancient Greek thinkers. While Euclid established modern geometry, there was little incentive to develop the skills of calculation. Greek number systems did not advance beyond those developed by the Babylonians.[12]

There was no pressing need to resolve the obvious ambiguities in Greek thought. The Greeks knew, for example, that a given load could more easily be moved to a given height with the help of an inclined plane than it could by being hauled straight up. But this fact did not bother them. By contrast, it was to torment Galileo. In his day, moving such loads was a significant engineering problem. That is why he came up with the concept of power, or rate of work, to explain what the Greeks only regarded as an interesting puzzle.[13]

The desire for fixed and eternal truths overwhelmed Greece, the more Greek society decayed. The degeneration of Greek society led to a decline in its science. The best material was produced early on. The Greeks' last great proposition was probably the heliocentric view of the solar system, put forward by Aristarchus around 250 BC. With Ptolemy's ugly mish-mash of epicycles, which emerged 400 years later, science was effectively frozen until Copernicus.

Trade, by the time of Copernicus, had catalysed the breakdown of Europe's old regional economy, and its twin bases: guilds in towns and the feudal system in rural areas.[14] As a result, the historian of science Stephen Mason suggests,

> during the sixteenth century the barrier between the craft and scholarly traditions, which up to that time had separated the mechanical from the liberal arts, began to break down. Guild secrecy faded out, craftsmen recording the lore of their tradition and assimilating some scholarly knowledge, while some scholars became interested in the experience and the methods of craftsmen.[15]

The deepening effect of trade and the demise of the classical guilds threw up the *mechanical philosophy*. Partly, that philosophy was a cipher, in ideas, for the mechanical methods being used in industry and war at the time. Partly, too, it represented a bold attempt to unify a wide range of disparate aspects of nature, even if the particular mechanisms behind these were not known. The physicist Robert Boyle pointed out that, though the truth of mechanical philosophy could not be proved, it was acceptable because it brought divergent phenomena into a coherent relationship.[16] Finally, the mechanical philosophy was also the outlook of practical scientists of the day. Frustrated by the previous tradition of always asking 'why?', they wanted instead to discover and catalogue things just as much as they wanted to understand their purpose.

In each new area addressed by science, the mechanical philosophy came to dominate. It penetrated some areas faster than others; as the American historian Charles Gillispie once explained: 'the attempt to answer the question "why?" carried the biologist much further into his science than it did the physicist'. But eventually, all came to believe in the superiority of the mechanical approach. As Gillispie went on to argue: in biology, it was perhaps more accurate to say that the question 'why?' became an obstacle much later.[17] All in all, the mechanical philosophy, crude as it often was, emerged as a key component of the buoyant post-Renaissance era.

Accounts of Science Which Downplay its Social Context

Many scientists and historians have highlighted factors other than society in their accounts of influences which have nurtured the development of science. The philosopher Paul Komesaroff has argued that many of the great breakthroughs in modern science first arose from a meditation about nature, especially in the realm of cosmology.[18] Kline also laid great emphasis on the development of ideas in their own terms:

> Our best knowledge of the behaviour of even those natural phenomena pervading our immediate environment has come from the contemplation of the heavens and not from the pursuit of practical problems.[19]

Even Einstein, whose consciousness of society and its effects on science was considerable, tended to see science as the outcome simply of previous science and inspiration. Responding to the question of why the Chinese had failed to develop modern science in the manner of Europe, he replied:

> The development of Western science has been based on two great achievements, the invention of the formal logical system (in Euclidean geometry) by the Greek philosophers, and the discovery of the possibility of finding out causal relationships by systematic experiment (at the Renaissance). In my opinion one has not to be astonished

that the Chinese sages have not made these steps. The astonishing thing is that these discoveries were made at all.[20]

To systematise Einstein's approach, the biologist Lewis Wolpert has, in his book *The Unnatural Nature of Science*, argued that science is entirely distinct from technology. According to Wolpert, technology arises through practical action, but modern science does not. The latter is profoundly alien to our way of looking at the world. It is the result of an inexplicable counter-intuitive leap. Wolpert concludes:

> the peculiar nature of science is responsible for the fact that, unlike technology or religion, science originated only once in history, in Greece.[21]

Here, the growth of science is related strictly to its organic, internal development.

Why the Social Still Comes First

To us, the 'internalist' account of science is one-sided. The practical interests of many of the world's pioneering scientists are now well-known, and the practical link between astronomy and navigation is obvious. Consider, for example, the title of the chair occupied by Adriaen Metius [1543–1626] at Leyden University, Holland: 'Professor ordinarus of mathematics, surveying, navigation, military engineering and astronomy'.[22]

Galileo, acknowledged as the first modern scientist, embodied the creative mixture of theory and practice which was characteristic of the sixteenth and seventeenth centuries. He tried to unify earthly and celestial mechanics – a task that was carried through by Newton – and was immersed in the technical problems of his day. Galileo's first job at Padua was as professor of physics and military engineering. His major work on mechanics, the *Discourse on Two New Sciences*, opens with a scene set in the Venetian arsenal, where Salviati, Galileo's mouthpiece, remarks: 'The constant activity which you

Venetians display in your famous arsenal suggests to the studious mind a large field for investigation, especially that part of the work which involves mechanics'.

The internalist explanation frequently skates over pragmatic preoccupations of individual scientists. But it also makes a more fundamental mistake. Internalist historians tend to reduce social progress to technology, and then discuss the relationship between technology and science. In fact, the growth of technology deserves to be explained itself, as does the rise of interest in sciences such as cosmology. There were, of course, aesthetic dimensions of the models of the heavens constructed by Copernicus and Galileo. But the *desire* to understand heavenly motion was not an aesthetic one.

The motivations of the scientific revolutionaries were fundamentally different from their Greek counterparts. For them, unlike the ancients, social advance opened up a whole new range of needs and possibilities. Social advance was a spur both to technological and to scientific development. Wolpert is right to argue that science did not significantly assist technology until the nineteenth century; but the development of practical problems gave scientists enormous food for thought. Thus, members of the Venetian arsenal knew how to fire their guns by rule of thumb – Galileo's theories did not assist them much. But Galileo certainly benefited from the intellectual atmosphere surrounding the arsenal. It coaxed him to understand the motion of projectiles more accurately than his predecessors.

Social progress also raised humanity's expectations of what was possible. It gave humanity confidence in the powers of reason; and that in itself was a significant factor in the development of modern science.

Einstein's point about the Chinese – that it was not astonishing that they did not re-invent the works of ancient Greece – is acutely made. Wolpert is also right to say that modern science is counter-intuitive. But these points only serve to highlight the debt science owes to a new society.

The birth of capitalism in Western Europe, and the social advance it brought about, helped to elevate the powers of human reason, and thus the spread of counter-intuitive thinking. Galileo was the first modern scientist precisely because he put reason above the immediate senses. As Salviati describes it in *A Dialogue Concerning the Two Chief World Systems*, using Aristarchus and Copernicus in Galileo's stead:

> I repeat, there is no limit to my astonishment when I reflect that Aristarchus and Copernicus were so able to make reason conquer sense that, in defiance of the latter, the former became mistress of their belief.[23]

The conquest of 'sense' by reason was by no means a purely intellectual affair. Much else had to be changing before the 'defiance' which Galileo wrote about and personally practised could ever hope to gather force.

Voltaire Spreads Newton's Ideas

The most famous populariser of Newton's work was the Frenchman Jean Francois Marie Arouet, better known as Voltaire [1694–1778]. For Voltaire, even the most famous statesmen and conquerors 'shrank until they seemed like figures in a rogue's gallery' when compared with Newton.[24] But Voltaire's life also shows the error in the view that the Scientific Revolution was the prime mover behind social change. Excited by science though he was, Voltaire was also engaged in a wider social project.

Humanism was the driving idea behind post-Renaissance Europe. 'Humanists' were, like Voltaire, enthusiastic about both the past achievements of humanity, and about what would prove possible in the future.[25] Science was, for them, an excellent but secondary means to more fundamental, and more human ends.

Exiled from Paris in 1726 for his criticism of the French government, Voltaire used science as much as he was influenced by it. Anglophile on account of the advanced stage reached by commerce and industry in Britain, Voltaire published, in 1733, his *Letters concerning the English Nation.* In it, in no fewer than four chapters, he mounted a detailed appreciation of Newton and his work. When Voltaire arranged for a French edition to appear in Paris in 1734, the work was condemned as 'likely to inspire a license of thought most dangerous to religious and civil order' and was banned.[26]

But it was not Newton's science, as such, that fascinated Voltaire and revolted his critics. Rather, Voltaire regarded Newton's work as an example of 'man's powers if he were set free'.[27] The fame Voltaire brought to Newton resided not in Newton's genius on its own, but in the broader way in which Newton's audacious outlook chimed with the changing mores of eighteenth-century Europe.

The Enlightenment Supersedes the Renaissance

Voltaire set the pattern for the thinkers who followed him. The French referred to the eighteenth century as the *siècle des lumieres,* the 'century of light'. The Enlightenment swept away the prejudices of the past and tried to establish, on a rational foundation, the study of humanity, society and nature. It also, as we saw in Chapter Two, implicitly opposed the naturalistic conception of humanity posited by the thinkers of the Middle Ages, separated human beings from nature, and upheld their ability to turn nature to their benefit.

Medieval society held the relation between humanity and nature to be fixed for all eternity. The Enlightenment, by contrast, felt that this relationship could change, but through a quantitative uncovering of the fixed mechanical laws.

Science was a battering-ram to be used to break down old prejudices and institutions. Condorcet described his peers as 'a class of men less concerned with discovering truth than with propagating it'. Such men, he continued, 'find their glory rather in destroying popular error than in pushing back the frontiers of knowledge'.[28] For progressives like Condorcet, then, the point about science was the impetus it gave to a wider vision of human advancement and perfectibility.

Looking back from the pinnacle of Enlightenment thought, Condorcet explained why Bacon, Galileo, and Descartes were the three founders of modern science and philosophy. Bacon had outlined the scientific method, but had done little real work. Then Galileo had led by example, in showing how nature should be explored. Finally, and above all, Descartes shook off tradition. After Descartes the path to human happiness was assured – or so thought Condorcet.

Despite their differing methods, what united Condorcet with Voltaire and other Enlightenment thinkers was their distaste for the authority of the past. The historian of science Richard Westfall has well brought out the significance of the leap they made over their predecessors:

> Renaissance Naturalism rested ultimately on the conviction that nature is a mystery which in its depth human reason can never plumb. Descartes' call for the abolition of wonder by understanding, on the other hand, voiced the confident conviction that nature contains no unfathomable mysteries, that she is wholly transparent to reason.[29]

There was, in fact, a certain ambiguity in Renaissance thought, a mixture of the old and the new. The new, the humanism that developed at the time, was still unsure of itself. It rejected the authority of received wisdom, but it did this by holding up the achievements of Ancient Greece.

With Bacon, Descartes and Galileo we have something fresh altogether – the forthright repulse of tradition. This attitude was summed up in the motto of the Royal Society: 'On the Word of No One'. The Ancients were still respected, but the new humanists put them in their place. As Santorio Santorio, professor of medicine at Padua from 1611 to 1624, argued: 'One must believe first in one's own senses and in experience, then in reasoning, and only in the third place in the authority of Hippocrates, of Galen, of Aristotle, and of other excellent philosophers'.[30]

The most aggressive attitude was probably articulated by

Bacon, who was freer than Santorio to speak his mind: Men, Bacon contended, had been kept back, as if by a kind of enchantment, from progress in the sciences. They had been impeded, he went on, 'by reverence for antiquity, by the authority of men accounted great in philosophy, and then by general consent'.[31] Thus, once scientists and philosophers gained confidence in their own powers of reasoning and observation, they grew openly contemptuous of the past. By the eighteenth century, as Norman Hampson has documented it,

> the men of the Enlightenment were virtually unanimous in decrying the Middle Ages. The entire period from the collapse of the Roman Empire in the West to the sixteenth century tended to be dismissed as one of poverty, oppression, ignorance and obscurantism.[32]

One example of the new posture is to be found in changed attitudes towards the Scriptures.

Galileo drew a distinction between the Scriptures and nature. He argued that a reading of the Scriptures could not afford to contradict the results of reason's investigation of nature; otherwise, the Scriptures themselves would be held up to ridicule. This was his point in the argument with the Vatican over Copernicus and his theory. Other thinkers were even more daring than Galileo. For them, rational investigation of nature was the only sure way to an understanding of God's work, because – unlike the Scriptures – nature had not been corrupted by humanity. In arguing this, they sought to overthrow the authority of the Catholic church.

Today, hundreds of years later, religiously-inspired thought still fights a rearguard action against science and the Enlightenment. Ironically it often does this by way of a polemic against the allegedly soulless, inhumane programme of science. Here Bryan Appleyard, the British journalist and author of *Understanding the Present: Science and the Soul of Modern Man*, is typical. Appleyard argues that the message of Galilean science is 'that we are nothing but trivial accidents'; and that 'each man must hope and believe what he can in the grim certainty that nobody and nothing will ever be able to tell him whether he is right or wrong'.[33]

It is true that Galilean science led to the conclusion that, in cosmic terms, humanity is nothing special. As Galileo's contemporary, Giordano Bruno, put it, 'Man is no more than an ant in the presence of the infinite'.[34] But Appleyard mistakes hostility to *tradition* for hostility to *humankind*. What people like Bruno rejected was not the realm of the human, but the teleology of Aristotelian and medieval thinkers – a framework which robbed humanity of any real freedom. To them, the cosmic insignificance of humanity was one thing; but a universe with no meaning at all was quite another. That is why their framework for science was not some deviant branch of technocratic philosophy, but a platform for liberation.

The success of reason in the scientific domain, and the universal achievements and aspirations of science, became the model for social reformers. In this sense Locke was inspired by the scientist Newton – but, like Voltaire, not so much by his science, as Butterfield would suggest, as by the potential which the same approach might offer with regard to social affairs. Locke believed that men were born equal, and that they were only made unequal by the iniquities of society. Through reason, there could ultimately be human happiness. The words of Condorcet sum up this optimistic vision best:

> The time will come when the sun will shine only on free men who know no master but their reason... How consoling for the philosopher who laments the errors, the crimes, the injustices which still pollute the earth and of which he is often the victim, is this view of the human race, emancipated from its shackles, released from the empire of fate and from that of the enemies of progress, advancing with a firm and sure step along the path of truth, virtue and happiness.[35]

By the time of Condorcet, reason proved an unstoppable force.

Take, for example, the discipline of history.

In the Middle Ages, there had been little enquiry into history. History, after all, was made by God. However, social change eventually led to a view of history which laid the accent on *progress*. In his *What is History?*, E. H. Carr cited the work of Edward Gibbon [1737–1794] as an example:

> Gibbon in eighteenth-century England saw history not as cyclical but as a triumphant advance: in his famous phrase, 'every age has increased, and still increases, the real wealth, the happiness, the knowledge and perhaps the virtue, of the human race'.[36]

In the light of this new outlook, there developed a perspective which emphasised the transience of human institutions; one which recognised that what existed might in time pass away. Some felt, indeed, that such institutions *ought* to pass away, if they did not match up to the needs of progress. Locke had, to an extent, anticipated these views with his relativistic conception of morality and politics. But Frenchmen like Condorcet and Henri Claude Saint-Simon [1760–1825] went further.

The Golden Age, Saint-Simon argued, 'lies ahead of us'. As the radical social commentator Frank Füredi notes, both Condorcet and Saint-Simon celebrated the power of human action to alter history. In their possession of what Füredi calls 'historical thinking', they looked back, but primarily to gain a sense of the future possibilities of change.[37]

Twentieth-Century Critics of the Enlightenment

It is the Enlightenment's scientifically-inspired vision of progress, and especially of human perfectibility, which so upsets many twentieth-century writers. Appleyard's vitriolic attack on Galileo might be a little strong for most – after all, even the Vatican has given the scientist a rehabilitation of sorts – but a more subtle critique has run throughout the twentieth century. That approach, briefly put, has been to obscure what the

Enlightenment was all about.

Alfred North Whitehead argued that the Enlightenment had been a useful corrective to the thinking that had gone before. But, he went on, it had gone too far in its opposition to metaphysics. Earlier, and seminally, at the beginning of the twentieth century, the French physicist and philosopher of science Pierre Duhem had made an ambitious attempt to defend the pre-Enlightenment past.[38]

Duhem tried to defend Aristotle. He also upheld the doctrines of the Catholic church. His argument was that Renaissance philosophy had built on, rather than opposed, the philosophy of the Middle Ages. Stripped of its explicit apology for Catholicism, this line of reasoning has had widespread influence.

Recent studies of sixteenth- and seventeenth-century science have concluded that few of the intellectual toilers of the Enlightenment were rational scientists at work for the good of humanity. One collection of essays, *Reappraisals of the Scientific Revolution*, opens by observing that 'while academic presses continued to provide undergraduates with the older histories, the same learned presses joined the specialist journals in producing highly focused studies that took root and began subtly to undermine the wall on which humpty-dumpty sat'.[39] In other words, modern science developed hand-in-glove with Medieval science.

Focusing on the psychology and practice of the individual scientists of old, today's revisionists proclaim a 'new historicism',[40] and hope decisively to overturn old nostrums about the scientific revolution.

Newton, we are reminded, spent as much time on alchemy as he did on science. Bacon did not do much science and even opposed Copernicus. Galileo aimed at glorying the creation of God, not the power of human reason. But in both their specifics, and in general, we believe, these fashionable criticisms miss the mark.

Of course there are Medieval elements in Bacon and Descartes. Yet these elements do not represent the essentials of their thought. Bacon and Descartes's essential contribution was to give a superior perspective on the development of knowledge. Before they put pen to paper, knowledge of reality was thought to be obtained through the *direct* awareness of the forms that constitute the essence of the objects of sense. Bacon and Descartes replaced this view with the idea that humanity *indirectly represents* the world and its structure through its sensory experience and its conceptualization. In doing so, they made a major advance, and one which must mark them off from earlier theory.[41]

What about Newton's alchemy and Galileo's religion? The real point here is simple. It would be a surprise if Galileo *had not* sought to glorify the work of God, and if Newton *had not* engaged in alchemy. To expect otherwise is rather naïve. Everyone carries the baggage of the past and the limitations of their age. How could it be otherwise when, for example, there was no knowledge of human evolution? What was irrational about believing in alchemy, when even the most rudimentary knowledge of chemistry was missing?

More to the point, Galileo and Newton were political conservatives. Numerous studies of the history of science and religion have shown that political radicalism was, and is, a more potent source of irreligion than Enlightenment rationalism or any other kind of rationalism pure and simple.[42] Thus Condorcet lost his faith in God only when the institutional role of the Catholic Church in French life, rather than abstract principles, persuaded him that he would need to make a change of heart.[43]

It is fun to trawl through the motivations of the world's historic scientists in a search for attitudes we only possess today. But, in its relentless hunt for detail, the exercise always runs the risk of losing the 'big picture' in science. An obsession with the personal idiosyncrasies of Enlightenment scientists can all too easily fail to apprehend the full significance of the scientific revolution.

In *The Destruction of Reason*, his monumental post-Second World War study on rationalism and irrationalism in human thought, Georg Lukács observed:

From Vico to Herder there runs a path which traces the extension, enrichment and consolidation of reason just as surely as the path taken by Descartes or Bacon leads in this direction. This gave rise to some very important differences, indeed antitheses, but all in all they were antitheses *within* a single camp fighting for a philosophy based upon the rationality of the world; nowhere do we find the abstract antithesis of rationalism and irrationalism.[44]

It was part of rationalism that its adherents would disagree. The revisionists make a mountain out of a molehill. The fact is that no dynamic revolution ever appears in pure, pristine and complete form, either in world-view or in deed.

Two Limitations to the Enlightenment

There are limits to Enlightenment thought. But such limits are not the ones that its contemporary critics tend to notice. There are two which should detain us here. The first is the crudeness of the Enlightenment vision of nature, and especially of humanity. Here we mean not the necessary limitations of the time, but rather a crudeness born of a structural defect in Enlightenment thought.

The second limitation of Enlightenment thought is its fragility. No sooner did it reach a peak in the opinions of Condorcet and Saint-Simon, than its supporters seemed to vanish. Where they did not vanish, as in the case of Saint-Simon, they tended to alter their earlier radical views.

To dwell on the first limitation: just as the initial dynamic behind science and Enlightenment thought was social change, so social change could circumscribe the two of them. The vision of nature and humanity developed in the Enlightenment represented the view of a new rising élite – the commercial and later industrial capitalist class. It reflected the hopes that élite had for its own society, together with the contempt it held for the old society from which it had sprung. But the limitations of its own society, and the threat it later felt from the dispossessed, came to exert a stifling influence on successors to the Enlightenment.

If the French positivist philosopher Auguste Comte [1798–1857] is taken as representative of early nineteenth-century views, then the fragility of the Enlightenment is all too clear. Comte was enthusiastic about science, but rejected quite emphatically the orientation to the future and historical change enunciated by Condorcet and the early Saint-Simon.[45]

Comte pioneered a vision of science in which it was still worthy of high esteem, but also robbed of any association with historical change and development. Comte developed a neutered vision of science and progress.

Comte's trajectory stretches out one that began, among many Enlightenment thinkers, the moment the French Revolution was complete. A profound swing in the direction of conservatism took place in élite thinking after 1789: indeed that year was a defining one for modern conservatism. Promises of human equality and progress soon came to be seen as highly problematic, because they only highlighted the failure of modern society to live up to those promises.

Comte's new synthesis of *science without reason* was developed in response to 1789. Before exploring the significance of shifts like this in more detail, a word more on the defects structural to Enlightenment thinking itself.[46]

We have argued that the mechanical view of nature developed by Descartes and others was a progressive step. Yet that point now needs to be qualified. What was progressive about the Cartesian framework was that it contained a conception of universal causality in nature. Any rational conception of natural law must contain this – for, without it, nature is made unknowable to humanity.

However, rationalism was not the sole influence leading to the widespread acceptance of the mechanical conception of nature. In his fascinating book *Pandemonium: The Coming of the Machine as Seen by Contemporary Observers*, Humphrey Jennings documents persuasively how *The Mechanization of the World Picture* gained real influence only once human labour was organised on a strictly regimented basis. In other words, it was the mechanical processes being developed in industry which inspired Enlightenment models of nature.[47] Jennings's idea is especially credible, given the close ties between the new élite and the scientists and philosophers of the day.

The mechanical models of life in particular that emerged were often quite crude. However, it was in explaining human nature that the crudities of Enlightenment thought were most clearly expressed. The gain that the Enlightenment made – and that many contemporary writers would like to eradicate – was to see humanity as superior to nature, and to develop a vision of progress based on humanity using nature for its own benefit.

Where, though, did these talents come from? That the Enlightenment theorists could not answer. For them to have done so would have required a conception of human progress that they lacked: namely, a view which recognised that human abilities and talents arose from social actions and interactions. Such a view did not exist among Enlightenment thinkers, except, somewhat incoherently, in the writings of Condorcet and Saint-Simon. The ascendant position of Enlightenment thinkers in the society of their day, and their allegiances to it, made it hard for them to envision change in society. As a result, the full consequences of human potential were beyond their comprehension.

Without a historical and sociological view of human nature, the Enlightenment thinkers oscillated between a religious view of human nature's innate characteristics, and a Lockean, *tabula rasa* approach. Neither of these perspectives did much to *explain* anything. They merely solved the riddle of human potential either by invoking God, or else by displacing it to another realm – that of humanity's interaction with its environment – without bothering to analyse exactly *how* this interaction could create the uniquely purposive properties that humanity possesses.

Burke, Comte and Mill: More Uncertainty about Science than is often Imagined

The French Revolution was the catalyst for a rapid shift in the intellectual climate in Europe – a shift in the direction of conservatism. In Britain, the statesman, essayist and philosopher Edmund Burke [1729–97] denounced the Revolution from the first. In doing so, he has a good claim to being the founder of modern conservatism. The tradition he began on science certainly deserves a look.

At first an isolated figure, Burke soon acquired a national prominence and a good deal of support. His main concern was the way in which the Revolution might upset propertied interests in Britain. To deal with this threat, he offered a direct counter to the forward, questioning attitude of Enlightenment; and, using Rousseau as his foil, he defended the power of ancient wisdom. In his *Letter to a Member of the National Assembly* (1791), he contrasted the wisdom that came from reading 'authors of sound antiquity' with the pernicious influence of Rousseau, 'the great professor and founding father of the *philosophy of vanity*'.[48]

Irving Zeitlin's book *Ideology and the Development of Sociological Theory* lists many other features of conservatism from the Burkean era, with particular reference to France. They included, among others: a hostility to the mechanical doctrine developed by Descartes; backing for stability and a view that meddling with society's order was dangerous; a positive assessment of the merits of irrationality; and support for old customs, institutions, and presuppositions as a vital cement with which to bind the status quo together. The archetypal French reactionary Joseph de Maistre [1753–1821] exemplified these trends:

Man's cradle must be surrounded by dogmas, and when his reason awakens, he must find all his opinions already made, at least those concerning his social behaviour. Nothing is more important for Man than prejudice.[49]

Such conservative ideas were not the only reaction to the French Revolution. Yet other responses, though not so clearly backward-looking, still rejected the optimism of the Enlightenment.

Auguste Comte, and John Stuart Mill [1806–73] were much more representative of Establishment thought in the nineteenth century than were Burke and others. Many of today's confusions about science and the Enlightenment stem from the misapprehension that these two thinkers in particular, and the schools they represented – respectively, modern sociology and utilitarianism – somehow continued the Enlightenment tradition.

In his *Age of Capital*, the historian Eric Hobsbawm argues, with some justification, that Comte and Mill founded their doctrines on the achievements of natural science. Of Comte's philosophy in particular, he writes:

> 'Positive' science, operating on objective and ascertainable facts, connected by rigid links of cause and effect, and producing uniform, invariant general 'laws' beyond query or wilful modification, was the master-key to the universe, and the nineteenth century possessed it.[50]

Hobsbawm is right to hint of the optimism of nineteenth-century scientists. But this was an optimism about natural science *in itself*. The link between the advance of natural science and the advance of human happiness, the link between science and reason, science and progress, which had so characterised Enlightenment thought had, by the nineteenth-century, been lost. To break the link was the very aim of Comte's positivism and Mill's utilitarianism.

Given the turmoil in France, it is not surprising that Comte's positivism had a more obviously conservative character to it than English utilitarianism. Comte's explicit aim was to unite *progress with order*. In fulfilling this goal, he rid progress of its dynamic character, since the preservation of order was his first concern. Comte always counterposed the preservation of order to the Enlightenment idea of human perfectibility. Daniel Pick captured this spirit of positivism excellently in his *Faces of Degeneration: A European Disorder, c1848 – c1918:*

> Nineteenth-century positivism held 'eighteenth-century theory' (the charges were often levelled in such vague and global terms) ideologically responsible for the excesses of the Revolution, partly for recklessly providing the masses with 'inflammatory' slogans, partly for its own conceptual failure to recognise the innate differences between individuals. The empirical fact of difference, so it was argued, made nonsense of catchwords like 'Equality'.[51]

Comte explicitly repudiated the idea that Reason could be applied to human history or society. The idea that humanity could change society according to goals it set itself was, wrote Comte, 'something which positivism must deny'. Comte then explained why, in words which are still laden with official and public support to this day. 'Goals', he maintained, 'in their very nature, are something that have not as yet been experienced'.[52] Because we have no knowledge of our destination, we should not set out for it.

The desire to separate science from social progress in fact united a wide range of nineteenth-century thinkers. It is a desire with which many thinkers today also feel comfortable.

The Nineteenth Century: Science Makes its Peace with Anti-Humanism

In Victorian Britain, science appeared to triumph. The biologist Thomas Huxley not only championed the cause of Darwinism, but also defeated Britain's bishops in debate and led the way to the secularisation of science. However, this is only one side of the story. For the secularisation of science did not amount to a challenge to the old authorities. Rather, a gradualist division of labour was worked out to the satisfaction of both sides. Neither would trespass on the other's patch. The status quo would be adjusted, not overthrown.

Many contemporary historians argue that the nineteenth century revived ideas of progress. But the views of Comte, Mill, and the 'Victorian compromise' of Huxley all show that this was not the case.[53] Rather than being a re-run of the Enlightenment, the nineteenth century anticipates attitudes prevalent in the twentieth. Science and technology continued to advance, very rapidly at times.[54] Scientists themselves were held in high regard.[55] But *science was now decoupled from Enlightenment optimism about improving the human condition.* This theme we take up in the next chapter.

In other ways, too, developments in the latter part of the nineteenth century prefigured twentieth-century attitudes. Oft-taken retreats from the material world into affairs spiritual contained more than a hint of antipathy to science. Even theoretical science was prone to lapse into idealism.

The Austrian physicist and philosopher Ernst Mach [1838–1916], and his use of Lagrangian dynamics, are a case in point. Around 1800, Joseph Louis Lagrange [1736–1813] had sought to unify all of mechanics through his discovery of a new mathematical formalism. This has since proved very powerful, and is now a part of undergraduate mathematics and physics. For Lagrange, it was designed to capture, in thought, the unity of material reality.

Writing at the end of the nineteenth century, however, Mach took a very different view of Lagrange's work. The formalism became, for him, a way to slip away from material reality into the realm of the formal itself. Through it, Mach started that pre-occupation with formalism which has been so prevalent in twentieth-century theoretical physics. This we examine in Chapter Seven.

Chapter Six:
Science and the Retreat from Reason

In the seventeenth and eighteenth centuries, optimism about science accompanied optimism about humanity. The rise of capitalism made anything seem possible. In the twentieth century, by contrast, recurrent tendencies toward economic decay and political exhaustion have created a vicious circle. Social progress now appears, and to a large extent is, highly problematic. In turn, the atmosphere surrounding social progress makes the direction and purpose of science uncertain. Then, finally, scientists' loss of bearings about the applications and benefits of their discipline, underpinned as it is by real economic and social constraints upon science, feeds back into wider society to deepen general fears and doubts about progress.

The Vicious Circle

The vicious circle which we have described has been evident throughout the twentieth century, but is at it tightest today. Yet the deepening gloom which today surrounds both science and progress should not distract from another fact: namely, that the twentieth century has witnessed a very rapid increase in the resourcing of and insights provided by natural science. Indeed what is so poignant about twentieth-century disenchantment with science is that it has emerged precisely when the conquests of natural science have been truly without precedent.

In 1951, Erwin Schrödinger wrote that it was 'doubtful whether the happiness of the human race has been enhanced by the technical and industrial developments that followed in the wake of rapidly progressing natural science'.[1] Two years later, Morris Kline was similarly pessimistic. He berated the Enlightenment:

> On the basis of the striking success achieved by Newtonian mathematics and science in the fields of astronomy and mechanics, the eighteenth-century intellectuals asserted the conviction that all of man's problems would soon be solved. Had these men known of the additional marvels science and mathematics were soon to reveal, they would have been even more unreserved, were that possible, in their expectations. It is now evident that these thinkers were indulging in unwarranted optimism.[2]

Like Schrödinger, Kline made his indictment at just the time when the application of science, in the form of atomic power, seemed to offer endless possibilities.

In fact this is the whole point. Technological advance, such as it has occurred, has tended to *reinforce* insecurities about science and about human judgement. It has rarely offset these. As the historian of science Stephen Toulmin suggests in his book *Cosmopolis*, the advance of science as knowledge and as technology over the past 400 years has not stopped contemporary culture and philosophy from turning its back on the Enlightenment.[3]

In coming to terms with the *simultaneous advance of science and retreat from reason*, the important thing is to explore all dimensions of the mutual interdependence of science and society. Few authors really attempt this. A brief example – that of Jonathan Piel, a leader-writer for *Scientific American* – is useful in highlighting methodological issues here.

Methodological Issues

A partisan of science, Piel contends that it shapes history:

Science is the force that prevents history from repeating itself. By creating knowledge with which to control nature or adapt to it, science breaks the pattern, turning what would be a circle into a spiral – usually, but not always, upward bound.

Science has the power to change both society and itself because answers always breed new questions.[4]

It is certainly true that history never repeats itself. But this is because society as a whole, including its scientific knowledge, moves on, rather than because science on its own has advanced. The idea that history never repeats itself also needs to be applied to the relationship between science and society. That relationship itself has changed over time.

Piel is right to say that the advance of science always poses new questions for society. However, scientific advance is also influenced by the questions that society asks of science. Furthermore, what is made of the questions asked by science depends on the state of society.

Our argument is that society has, over the course of the twentieth century, shown a decline in its power to realise the full potential of science. It is this fact which explains why, on its own, the overall advance of science fails to enthuse the world as it might, and certainly fails to generate a sense of progress. Again, too, there is a circular dynamic at work here: once science slows up in relation to what it *could* do, society loses faith in it. In turn, that can only disorientate the international scientific community. Then the disorientation of scientists itself impedes the advance of science.

The vicious circle to which we have referred has, in sum, the following foundation and characteristics: *despite the advance of science*, social progress has become a fraught affair, and is seen as such. As a result, the international scientific community fragments into a group of individuals with little common purpose. This circumstance then reinforces the fact that today's scientific breakthroughs, though more conspicuous than ever, are rarely generalised in ways which are fully relevant to humankind.

Science and the Retreat from Reason

It is important to determine how the *combined advance of science and retreat from reason* have come about. Economic stagnation and military conflicts have certainly played their part in reducing humanity's expectations of science and of the future. But it is the more and more obvious redundancy of political institutions and movements which has done most to discredit the scientific cause – and the cause of progress.

Looking back over the twentieth century, it is clear that the traditional political categories 'right' and 'left' have come to share many things in common.

In the late eighteenth and early nineteenth centuries, the Enlightenment bequeathed upon Europe a momentous struggle between the forces of absolutism and those of reform. Out of the class struggles of the first part of the twentieth century emerged the modern divide between right and left. But the end of the Cold War revealed that the apparent polarities of the past 50 years served to hide a convergence of left and right. Today both traditions are beset by a lack of public respect. Both are unwilling, if not unable, to offer much of a vision for the future. Both are, indeed, pretty exhausted.

Right and left now agree that there is no alternative to the market system. But, and this is symptomatic of their exhaustion, both now recognise that the market can neither cohere society, nor offer any hope for the future. For the right, this amounts to disenchantment with its own society. For the left, it represents the end-point of decline. Starting after 1848, and reinforced by the events of 1917, the historic goal of the left was an alternative to the market. The end of this project marks the end of the left, and indeed of any meaningful left/right divide.

Both right and left now share a profound distrust toward millions of human beings – whether these people are members of an unsavoury 'underclass' or an alien nation.[5] Both agree that there is a threat to the planet posed by the growth of human population. The two sides also agree that economic growth must not be allowed to fuel inflation, or to threaten the environment.

Altogether, there is a desire to limit human ambition. The management guru Charles Handy has captured the mood in his *The Empty Raincoat: Making Sense of the Future* (1994). For Handy, 'original sin is the price we pay for our humanity'. Faced with a world in which 'the Theory of Complexity has been added to the Theory of Chaos', mankind has reached a moment where there are 'few great causes or crusades any more'.[6]

The sense of terminus in society today is, in fact, only a culmination of twentieth-century trends as a whole. On the right, premonitions of decline began early on. The traumas of the First and the Second World Wars compounded nervousness. Even though the subsequent Cold War assured an identity for the right after 1945, fascism proved to be an episode which it could not easily shrug off.

Today, the right's victory over the left in the Cold War has proved a pyhrric one. If the 1990s were ushered in by the final triumph of market forces as a social system, how is it that the economic 'recovery' which has succeeded the recession of 1989–1991 still feels so fictitious to so many people in the West?

As they are currently articulated, the outright defenders of market forces face an irreversible loss of momentum. In turn, the right's paralysis can only underline its historic preference for matters financial and spiritual over matters scientific. The right can never, under any circumstances, write scientists a blank cheque. Pockets are shallow and research goals must not be too abstract or ambitious. Science programmes must everywhere deliver value for money.

At the same time, of course, the right says that science is no substitute for personal and moral responsibility. Science is said to be inadequate. It has nothing to contribute to the soul, spirit, heart or to God. Science must therefore be tolerated but not especially venerated. It must know its place.

Left-wing thought has gone through a different trajectory, but has ended up at a similar finishing-point. The left began life as a champion of science and progress. In particular, the left used to indict capitalist society for the economic, commercial and international barriers it threw up to the full utilization of science. During the era of Rachel Carson's *Silent Spring* (1962), however, liberal critics of industrial society began to attack it not for its unfulfilled potential, but for being too hasty and reckless in its application of science and technology. From Vance Packard's *The Waste Makers* through to today, the radical critique of capitalism was and remains that it was producing and consuming too much, not too little. The problem was seen and remains seen as one of *waste*, and in particular humanity's seemingly deep-seated tendency to go about *laying waste to the natural environment*.

By 1995, the left's main theme was the need for human beings not just to exercise rights, but also to fulfil civic responsibilities. The left advocated environmental restraint. It also favoured more state regulation in every field – science included.

The left can no longer even remember its old critiques of contemporary society in the name of science and progress. On the surface, the left still favours technology and the scientific enterprise. But beneath the surface, the left harbours a deep tiredness about the prospects of society, and its own prospects as a social force. It too believes that there is no such thing as a free lunch in science. Like the right, it wants value for money in science, and is highly suspicious of both real and imagined perversions of science. Yet this is only a piece of a larger preference, on the contemporary left, for change on a cautious scale, fiscal rectitude, and market forces.

Today there is a clear disjunction between scientific advance on the one hand, and cramped right and left perspectives about society on the other. And that disjunction has consequences, for both science and society. These consequences, in turn, can be damaging – as we explore at the end of this chapter. In the meantime, we turn to the individual evolution, and mutual interaction, over the twentieth century, of attitudes toward science and social progress. We begin by looking at conservative and liberal traditions before and after 1914.

On the Right: War Confirms Disillusionment with Progress and Science

The years either side of 1914 were beset by social and international instability. During those years both liberal and conservative thinkers began to lose faith in their own society. Once again, E. H. Carr documented the shift of mood with percipience:

> Between the middle of the last century and 1914 it was scarcely possible for a British historian to conceive of historical change except as change for the better. In the 1920s, we moved into a period in which change was beginning to be associated with fear for the future, and could be thought of as change for the worse – a period of the rebirth of conservative thinking.[7]

Carr went on to point out that, in response to the new realities, another British historian, Arnold Toynbee, had made 'a desperate attempt to replace a linear view of history by a cyclical theory – the characteristic ideology of a society in decline'. Since this failed attempt, Carr then concluded, British historians had, for the most part, 'been content to throw in their hands and declare that there is no pattern to history at all'.[8]

For Carr the watershed of world war led to a reemergence of conservativism. Another way of putting this is that the old nineteenth-century distinctions between liberals and conservatives lost their force. Both traditions were sapped of vigour and confidence, and, in Britain, the Whig and Liberal vision of social progress endured its final collapse when, in the First World War, Liberal England became conservative in all but name. However, this shift in a conservative direction did not lead to an explicit British reaction against science. In Britain, the élite did not suffer such a crisis of faith as that experienced by its continental counterparts.

On the continent, the conservative reaction to science and to progress was forthright. It emphasised the power of the irrational over the rational. Among the middle classes, during the period 1871–1918, science lost out as the power of the irrational expanded. A foreboding grew that dark, unfathomable forces were at work in the lower reaches of society. This was the era of fear of the crowd (in French *foule*, in German *masse*, both renderings which have political associations which the English 'crowd' lacks).[9] In general there was an upsurge in the vitalistic ideas of the French philosopher Henri Bergson and others, and a disparaging of scientific and technological achievement.

In France, the shift of mood between 1871 and 1918 is striking. Defeat at the hands of Germany in 1871, coincident with the Paris Commune, unhinged a generation of French conservatives. The élite did not, however, turn away from science and social advance immediately. As Robert A. Nye points out in his excellent *The Origins of Crowd Psychology: Gustave LeBon and the Crisis of Mass Democracy in the Third Republic*, the immediate perception was that

> much of France's failure was the failure of her scientific education. For France's rebirth, as Flaubert wrote in a post-war letter to George Sand, 'she must pass from inspiration to science'. Even the sceptical agreed that the 'new idol' of science and scientific method was a major hope for regeneration.[10]

Only when France, aided by French science, eventually failed to achieve the hoped-for regeneration, did a large and influential section of the French élite turn toward outright political reaction – against republicanism, and against science. Nye illustrates the shift through the object of his study, the reactionary French theorist of crowd behaviour, Gustave LeBon [1841–1931].

LeBon, after 1871, hated large groups of proletarians and the implicit threat to order which they represented. For several decades after 1871, he argued that scientific advance would help deliver some of the tools needed to keep the crowd in its place. However, as Nye observes, 'as the first decade of the [twentieth] century drew to a close... his commitment to the "solidarity" of nationalism increased'.[11]

LeBon's attitude to the values of the Republic and of science grew more hostile. And in this shift, he mirrored that made by thousands of his compatriots:

> In the face of the apparent failure of a revival of material culture, many Frenchmen concluded with LeBon that only a spiritual and moral renewal could regain for France the position in the world she had enjoyed half a century earlier.[12]

Moral uplift, not scientific ascent, became the watchword of French conservatism.

In Germany, the nation's defeat in war led, on the right, to a mass outbreak of pessimism about both society's fate and that of science. In particular, Oswald Spengler's bestseller *The Decline of the West* (1918) found a ready audience among a generation shellshocked by the collapse of the Western Front. Spengler believed that there had been, in the West, a victory for scientific values. This, he went on, was directly associated with the decline of Western civilisation. Spengler held that, at the highpoint of a civilisation, the arts flourish. By contrast, culture decayed whenever the rule of the sciences went unimpeded. For Spengler, Germany, the land of Goethe, had allowed itself to worship science, and to grow infatuated with technology.

In his influential *The Reorientation of European Social Thought 1890–1930*, the intellectual historian H. Stuart Hughes sums up his chosen period, which supplied the overarching context for thinkers like Le Bon and Spengler, as a 'rebellion against positivism' – or at least against positivism 'in a vulgarised sense'. He writes: 'both in the "lower" and "higher" levels of intel-lectual activity, doubts arose as to the reigning philosophy of the upper middle class – the self satisfied cult of material progress'.[13]

In our view, the irrational thinkers H. Stuart Hughes writes about were actually more engaged in a rebellion against the Enlightenment than in one against positivism. This was so, even though positivism had already supplanted the ideas of Enlightenment well before 1914.

All the continental forces of philosophical reaction lambasted the Enlightenment for its faith in progress. At the same time, the right had no serious quarrel with science as a body of techniques. For this reason, we regard the irrational tradition as a variant of positivism, rather than as a reaction against it.

This is indeed a crucial point. That modern society has a profound *need* for science is shown by the fact that no liberal, conservative or 'irrational' thinker of any significance after 1914 actually opposed natural science as a body of techniques. A vivid example of that fact is provided by the practice of the Nazi regime.

Many writers have highlighted how Germany's Third Reich rejected scientific values. After 1945 Samuel Goudsmit, who was scientific director of the American intelligence investigation into German efforts to produce an atomic bomb during the war, claimed that, in part, the Germans had failed to produce a bomb because of the anti-scientific postures of the Nazi regime. The point seems reasonable enough. After all, Adolf Hitler himself said:

> We stand at the end of the Age of Reason... A new era of the magical explanation of the world is rising, an explanation based on will rather than knowledge. There is no truth, in either the moral or the scientific sense... Science is a social phenomenon, and like all those, is limited by the usefulness or harm it causes. With the slogan of objective science the professoriat only wanted to free itself from the necessary supervision of the state.[14]

It is also well-known that fascism in Germany opposed much of modern science as 'Jewish Physics'. Yet even the mystically-inclined Nazis realised that gaining an atomic bomb meant using modern science. It was not antipathy to science which prevented Hitler's forces from making a bomb, but rather the low priority afforded to atomic weapons as a project.[15]

The undeniable prowess of science forced even the National Socialists to embrace it. In the same spirit, irrationalists like Bergson tried to couch their ideas in scientific phraseology. As the French conservative Ernest Renan conceded:

> Today war, mechanics, industry require science, so much that even the persons the most hostile to the scientific spirit are obliged to learn mathematics, physics and chemistry. In all ways, the sovereignty of science is being imposed, even on its critics.[16]

Summing up the right and its attitude to society and science in the opening decades of the twentieth century, we believe that the First World War deepened unease about both. The 'masses' were here to stay, and so was science and its influence; but conservatism viewed neither development with equanimity.

On the Left: Caution about the Benefits of Science

Like that central figure of Cold War dissent in American sociology, C. Wright Mills, H. Stuart Hughes always likes to draw contrasts between the 'irrationalist' thinkers who form the subject of his book, and both liberal and mainstream-conservative thinkers. According to Hughes, men like Max Weber succeeded in keeping the Enlightenment tradition alive in the face of the irrationalist onslaught. For us, however, examining the key liberal thinkers in this period suggests a different conclusion.

'Fear of the future', about which Carr wrote, pervaded the left just as it did the right. Moreover, liberals joined irrationalist thinkers in their growing despair about science and Enlightenment notions of progress. In sociology, psychology, and philosophy respectively, Max Weber, Sigmund Freud and Ludwig Wittgenstein all illustrate this trend.

Weber's well-known distinction between facts and values led him to counterpose science to culture and society. It was 'one thing to state facts, to determine mathematical or logical relations or the internal structure of cultural values'; it was quite another 'to answer questions of the *value* of culture and its individual contents.'[17] Thus science was all very well, but had no special link to progress. Indeed, science tended to march on at the expense of human feelings.

Weber's bipolar conception of science and culture also had a historical side to it. Looking at art and culture, Weber felt that their peak lay in the past. For him the twentieth-century world was *disenchanted*, as the relentless march of urbanisation, organisation, bureaucracy and technology robbed it of charm. To his credit, Weber believed that the desire to recreate the past was a childish one. Nevertheless, he mourned its passing.

Weber has proved one of the key influences on twentieth-century thinking. This applies to thinking about science as much as anything else. After Weber, many saw, and still see, science as the amoral steamroller of modernity, whose ruthless crushing of community and tradition is as sad as it is inevitable.

Freud too counterposed the advance of science to social advance. He focused on the clash between scientific achievements and psychological well-being. Discussing the former in his famous *Civilization and its Discontents* (1930), Freud wrote:

> Men are proud of these achievements, and have a right to be. But they seem to have observed that this newly-won power over space and time, this subjugation of the forces of nature, which is the fulfilment of a longing that goes back thousands of years, has not increased the amount of pleasurable satisfaction which they may expect from life and has not made them feel happier.[18]

Freud was here not just commenting on the manifest inability of his own day's science to cure the Depression psychology of the 1930s. In his writings, it is also evident that Freud decried science's ability *ever* to rid humanity of neuroses, psychoses and other features of mental ill-health.

Especially in his early work, Ludwig Wittgenstein was keener about the benefits of logical and scientific thinking than was

Freud. But as his recent biographer Ray Monk relates, Wittgenstein believed celebration of science to be a symptom, perhaps even a cause, of that decline of Western culture which he came to feel so keenly. In Wittgenstein's more morbid moments after the Second World War, he even thought that atomic destruction might be no bad thing: the Bomb, he wrote, offered 'a prospect of the end, a destruction, of an evil – our disgusting soup of water science. And certainly that's not an unpleasant thought'.[19]

In the 1930s, Wittgenstein investigated the possibility of living in Russia. This was a serious decision. His desire to move East did not express support for the Soviet system, as much as dismay about the course taken by Western society. As Monk convincingly argues:

> Wittgenstein's reasons for wanting to live in Russia, both the 'bad and even childish' reasons and the 'deep and even good' ones, had much to do, I think, with his desire to dissociate himself from the old men of the West, and from the disintegrating and decaying Culture of Western Europe.[20]

An estrangement from science, progress, and the West enveloped European intellectuals like Weber, Freud and Wittgenstein. But this took place, we should again note, at a time when science and technology had become organised as never before. We have seen how Planck, Einstein and others revolutionised world science. The First World War also ensured that nation states went about organising the national scientific effort in a systematic way. Then, after the war, Einstein acquired cult status. He was asked to put on his own three-week show at the London Palladium, women fainted in his presence, and young girls mobbed him in Geneva.

But the high profile of science and scientists did not signify a culture dominated by notions of progress akin to those of the Enlightenment. Einstein captured the popular imagination because of the brilliant contrast he offered to the start of a much darker era – one in which humanity seemed to have lost control of the world. Science was thus celebrated in ways which tended to counterpose it to society. Removed from the political turbulence and chaos of inter-war Europe, science was put on a pedestal. For many it inspired not rationality, but escapism.

The twentieth century has seen rapid advances in science and technology. But, as both conservatives and liberals went into and came out of the First World war with a shudder, politics acquired a sense of exhaustion. The result was that the brilliant insights and practical leaps forward achieved by pre-war and inter-war science were often stripped of any real link with the project of bettering the human condition. Of course, New York could still play host, as it did in 1939, to a World's Fair stuffed full of technological marvels. These, forward-looking designers and architects envisaged as happening 'in our lifetime'. But for the immediate present, and for the first time, life looked too bleak for science to be a vehicle for very much more than fantasies.

US Science in the Post-Manhattan Era: How the Cold War Generated Scientism

It was in America after the Second World War that hopes about science were highest. However, as we showed in Chapter One, those hopes did not translate into ideals about the future. At the inception of the space race, Edward Teller, renowned as the 'father of the H-Bomb', observed:

> Today I do not read science fiction. My tastes did not change. Science fiction did. Reflecting the general attitude, the stories used to say 'How wonderful'. Now they say 'How horrible'.[21]

Two world wars and an intervening economic slump had engendered an obsession with 'security'. In the Cold War this was articulated as a problem with the Soviets; but in fact America's insecurities related to the future in all its forms. They were insecurities which technological advance and economic expansion after the war did not fully overcome.

In the first place, long-established American doctrines of white supremacy were discredited by the Nazi experience in Germany.[22] Then China was lost, and eventually Vietnam went the same way. The land of Vannevar Bush could still rule much of the world by force, the dollar and technology; but it was by no means confident that it knew the way ahead. The forward march of knowledge after 1945 was very different in ethos and spirit from that which animated America's Founding Fathers, not to mention the pioneers of the French Enlightenment. After the Manhattan Project, growth in scientific and technical knowledge was never again to be accompanied by jubilation about the potential of human beings. Quite the contrary.[23]

The American élite's endorsement of science emerged with the political glacier of the Cold War. In this climate, the 'new priesthood' of scientists which grew up around and long survived Manhattan did, it is true, stimulate the development of science- and technology-driven utopias. However, what distinguished the dreams of post-Manhattan science was the way in which they were shaped by, and mirrored, the world of Cold War confrontation.

The future which was mapped for both science and technology after 1945, and the everyday practice of these disciplines, often depended on some of the more reductionist methods of natural science. Even more inappropriately, these methods were also applied to the realm of personal and social affairs. In the place of a humanistic framework, there arose the anti-human straitjacket of what we term *scientism*.

RAND, the company that became synonymous with the term 'think-tank', best illustrates these trends.[24] RAND, a corruption of 'Research And Development', began life at the Douglas Aircraft Corporation. In 1948, with the blessing of the US Air Force and loan guarantees from the Ford Foundation, it then cut loose and became the RAND Corporation. At its inception, RAND viewed itself as 'fundamentally interested in and devoted to what can broadly be called the rational life'.[25] But

RAND's thinking was neither so broad nor so rational as it made out.

For RAND, 'rationality' meant that society could be analysed using the tools of natural science. RAND's was a highly technocratic vision, which sought to reduce politics to a purely quantitative discipline, and which had no qualms about applying the most esoteric mathematical tools to the calibration of nuclear 'collateral damage'.

In both political and military affairs, the goals of America's élite and the structures of society were left unquestioned. The role of rationality was confined to the study of resource and conflict management. In turn, the management of the Cold War was, for RAND, very much about American victory. Victory was necessary against the Soviets even if their cities would have to be annihilated; indeed, even at the expense of American cities, too.

The militaristic thinking of RAND was expressed well by two of its leading thinkers: the Hungarian genius John von Neumann, and Herman Kahn. Von Neumann worked on the Manhattan Project during the war. His colleagues in that enterprise came to have regrets about the use of the Bomb, but von Neumann himself did not. A leading advocate of a nuclear first-strike against the Soviet Union, he once memorably argued:

If you say why not bomb them tomorrow, I say why not today? If you say today at 5 o'clock, I say why not one o'clock?[26]

No wonder it is alleged that, in 1963, von Neumann turned up as one of the real-life models for Stanley Kubrick's Armageddon on film, *Dr Strangelove*.

Herman Kahn was just as sinister. Kahn worked as an employee at RAND in the 1950s before moving on to found the Hudson Institute in 1961. In the early 1960s he published two books which made his name: *On Thermonuclear War* and *Thinking the Unthinkable*. One of the 'unthinkables' was that Americans would have to suffer a thermonuclear strike in the cause of the Cold War.

RAND soon took advantage of its ex-employee's thinking. It devised plans on how to ration access to nuclear fallout shelters. One of RAND's suggestions was to let children into such shelters first, and then exclude their parents – in the hope that the grown-ups would sacrifice themselves fighting off other adults desperate to enter places of safety.

RAND and Kahn, no doubt, had bunkers already allocated. Indeed, the repressive implications of RAND's policies and of Cold War scientism in general were rarely avoided in the literature of the era. The destructive uses to which science and technology could be and were put by the US military were fully documented. By the second half of the 1960s, they were evident to many, and especially to the radical protest movements of that time. Scientism thus created opponents for itself.

The attitudes of those opposition movements toward science and progress have echoed down the years, and are now very influential. Before we examine those attitudes, we need to provide a little background on the typical radical approach toward science and progress as it existed in the years before the '60s generation made its mark.

Old Left Scientism Meets its Demise

The Old Left, schooled in the late nineteenth- and early twentieth-century Socialist and Communist parties, approved of science and, for three decades after the Wall Street Crash of 1929, tended to see Moscow as broadly more adept at making the most of science than Washington. Typical of this tradition was the British scientist J. D. Bernal. In 1954, he wrote:

> The transformation of nature, along the lines indicated by the biological sciences, will be undertaken with the use of heavy machinery, including possibly atomic energy. All the river basins of the world can be brought under control, providing ample power, abolishing floods, droughts, and destructive soil erosion, and widely extending the areas of

cultivation and stock raising... Beyond this lie possibilities of further extending the productive zone of the world to cover present deserts and mountain wastes and making full use of the resources of the seas, and beyond that again lie the possibilities of microbiological and photochemical production of food.[27]

For Bernal, science made anything possible.

In France, the physicist and communist party member Frédéric Joliot-Curie also saw science in a wholly positive light. Writing just days after the bombing of Hiroshima and Nagasaki, he argued: 'I am personally convinced that, despite the feelings aroused by the application of atomic energy to destructive ends, it will be of inestimable service to mankind in Peacetime'.[28] Ten years later, Joliot-Curie saluted atomic energy technologies once again. They were, he said,

forces liberated by Man, and Man has complete power to direct their use exclusively for peaceful ends. The situation would be quite different if we had to deal with a brutal threat from natural forces such as that offered by the forecast of an imminent collision between our planet and an immense meteorite.[29]

The tracts of Western communist parties after the war often read like this. Gradually, they dropped the old proposition that the West could not develop or implement science to the full. Instead, in a lurch to scientism almost the equal of the RAND Corporation, they concluded that the full implementation of science, together with East-West détente, would cure all ills. 'Peaceful coexistence', underpinned by an unpleasant but perhaps irreversible nuclear balance of terror, became the world view of the old Stalinists of the West, and especially of Western Europe.

In the 1960s, the frostiest years of the Cold War eventually gave way to a growing rapprochement between the USA and the USSR. In the process, critics like Bernal and Joliot-Curie came to terms with a system they had earlier despised. There was more, too: because the Old Left reduced progress simply to the advance of science, it saw in the economic and science booms of the early 1960s reason enough to join in government initiatives on science and technology.

For example, Bernal became an advisor to Harold Wilson's government of 1964–1970. He took at face value Wilson's famous speech to the Labour Party conference in 1963, in which the Labour leader declared:

We are redefining and we are restating our socialism in terms of the scientific revolution... The Britain that is going to be forged in the white heat of this revolution is no place for the restrictive practices or outdated methods on either side of industry.

Old Left figures like Bernal now made reform of the existing administration of science their goal. They jettisoned once and for all critiques of capitalism's ability to maximise the benefits of science. But at just this moment, élite circles in the West proved to have a short attention-span when the issue of reform was practically posed.

Government attitudes toward science grew more critical on both sides of the Atlantic. The Apollo programme eventually wound down; eventually too, Pentagon technologists went into hiding after the end of the Vietnam war. Meanwhile, a number of collaborative science projects connected with the then European Economic Community proved prolonged in execution, expensive and, many argued at the time, ultimately ill-conceived. EURATOM, the European Launcher Development Organisation, the European Space Research Organisation... all these faced a growing body of critics. Meanwhile national science and technology projects in Europe fared little better.

They too encountered official hostility. Typical of the mood that followed was the comment of a Secretary of State for Science and Education in Britain's 1974–79 Labour government. 'For scientists', Shirley Williams declared, 'the party is over.'[30]

For those ageing scientists who had once idolised Stalin, the party had ended long before. Their uncritical endorsements of science found fewer and fewer listeners. Eventually, radical scientism disappeared from the West, as recessionary conditions and a wider shift in mainstream attitudes took their toll.

From Pre-War Germany to Post-War America: The New Left Attacks Science Per Se

In reaction both to RAND-style militarism and Stalinism, a number of alternative radical traditions sprung up in the late 1950s and 1960s. In Britain, the 'New Left', a characteristic member of which was the historian E. P. Thompson, came out in opposition to both East and West. It advocated a broadly humanistic and forward-looking socialism. A quite different movement, which has also come to be known as the 'New Left', emerged a little later on. As a movement, this New Left was more influential internationally: it embraced both political protest in Germany, and the American 'Counter Culture'. Its leading intellectual was the German Herbert Marcuse. What marked this New Left out as a distinct political tradition and one of lasting significance was that, unlike its British namesake, it rejected Enlightenment notions of science and progress. In combination with the emerging Green movement, this New Left helped establish a seemingly radical argument for suspicion of science. Having examined Old Left scientism, we now take a look at the radical rejection of progress.

Like Old Left scientism, this second strand in radical post-Manhattan opinion on science can also be traced back to the 1930s. Initially a minority influence, it eventually came to supplant the Old Left's scientistic world-view.

In Germany before the Second World War, Theodore Adorno and Max Horkheimer were among the most influential members of that tradition of left-wing criticism known as the Frankfurt School. Then, during wartime exile in America, Adorno and

Horkheimer wrote their celebrated, if little read, *Dialectic of Enlightenment*.[31] This became the founding text of New Left opposition - not just to the path taken by science in capitalist society, but to the Enlightenment's project of progress itself.

In their book, Adorno and Horkheimer put forward two ideas which later turned out to have a seminal influence. First, they held that modern society had come to grief because it was founded on the realisation of Enlightenment ideas. Second, they argued that the growth of human dominion over nature was dehumanising. For Adorno and Horkheimer, the 'fully enlightened earth' represented 'disaster triumphant'.[32]

Control over nature was wrong. The Enlightenment had turned nature into 'mere objectivity' for humankind. As a result, human beings were forced to

pay for the increase of their power with alienation from that over which they exercise their power. *Enlightenment behaves towards things as a dictator towards men.*[33]

Adorno and Horkheimer did not stop there. For them, the hubris bound up with humanity's attitude toward nature helped foster dictatorial attitudes toward subordinate classes and countries.

The emergence of fascism in Germany confirmed Adorno and Horkheimer in their view. The 'unconditional realism' of the modern condition, they wrote, 'culminates in fascism'. That realism, founded in the Enlightenment, was for them 'a special case of paranoic delusion which dehumanises nature and finally... nations themselves'.[34]

Adorno and Horkheimer wrote at a time when Hitler had smashed the German left, once the strongest socialist movement in inter-war Europe. So it was not hard for them to draw the bleak inferences they did. But as our earlier quotation from Hitler showed, the Nazis denied 'realism' and rationality. For Adorno and Horkheimer to regard fascism as embodying the Enlightenment was an understandable but nevertheless ludicrous mistake. Unable to see an alternative to the way society was, the authors elided the destruction which twentieth-century society had wrought into the whole endeavour of progress itself.

In her polemic against the use of insecticides, *Silent Spring*, the American marine biologist Rachel Carson arrived at a similar elision. Arguing that life, including insect life, was a miracle 'beyond our comprehension', she contended that it deserved reverence and humility from human beings.[35] She went on:

> The 'control of nature' is a phrase conceived in arrogance, born of the Neanderthal age of biology and society... It is our alarming misfortune that so primitive a science has armed itself with the most modern and terrible weapons, and that in turning them against the insects it has also turned them against the earth.[36]

For Carson the Enlightenment drive to manipulate the world through science deserved the same stigma which Adorno and Horkheimer heaped on it during the war. Such a quest was, for her, 'Neanderthal'.

Carson was much more widely read in her time than Adorno and Horkheimer were in theirs. As a result, her attack on science as conceited and arrogant was unique in its achievement. For the first time in centuries, this ex-official of the US Bureau of Fisheries had scored a hit, not simply against progress, but at the most basic right of science to change the outside world.

Carson was not at all a member of the New Left. But the enthusiastic reception granted her ideas by the Kennedy administration ensured that a growing environmental movement in America soon came under her spell. Nearby, the New Left, as it developed on American campuses, was profoundly assisted by the climate set by Carson. In terms of ideas, however, it was also animated by an old Frankfurt School compatriot of Adorno and Horkheimer: Herbert Marcuse.

Marcuse's book *One-Dimensional Man* was first published in 1964. In it, Marcuse contended that people and nature were dominated by a suffocating, one-dimensional rationality which recognised neither human nor natural variety: 'the world', he protested, had 'become the stuff of total administration'. There seemed to be no way out, because 'the web of domination has become the web of reason itself, and this society is fatally entangled in it'.[37]

Any person interested in breaking out of such a world, he argued, must reject Old Left notions about the neutrality of science and technology and the merits of control of nature. Why? Because 'with the growth in the technological conquest of nature', argued Marcuse, 'grows the conquest of man by man'.[38] For him, not just RAND but the complete and 'spurious' notion that technology was neutral merely concealed the interests of the American élite and 'the politics of domination.'[39]

Marcuse claimed that, from the Enlightenment onwards, modern science and technology were actually *instrumental constructions*. Science did not reflect the way nature was, but rather what élites wanted it to be. In the same way, technology reflected élite aspirations; by no means was it simply a neutral set of tools which could be used for good or bad according to who was wielding it. In a striking passage, Marcuse wrote:

The principles of modern science were a *priori* structured in such a way that they could serve as conceptual instruments for a universe of self-propelling, productive control; theoretical operationalism came to correspond to practical operationalism. The scientific method [which] led to the ever-more-effective domination of nature thus came to provide the pure concepts as well as the instrumentalities for the ever-more-effective domination of man by man *through* the domination of nature... Today, domination perpetuates and extends itself not only through technology but *as* technology, and the latter provides the great legitimation of the expanding political power, which absorbs all spheres of culture.[40]

Yet if, as Marcuse claimed, modern science represented élite prejudice rather than an understanding of nature, it should have been possible to *show* this through providing a *better* natural science. This Marcuse neither did nor called for. Conversely, if modern technology merely amounted to domination, Marcuse's analysis implied a return to pre-Modern or pre-Enlightenment technologies. Again, Marcuse did not call for this, even if, in the abortive rural communes of the 1960s, many of his youthful American followers essayed a passage back to agricultural modes of life.

Marcuse rejected the Enlightenment. But he offered no clear alternative to it. It was his own conclusions which were 'one-dimensional'. In place of the 'domination' of rationalism, he substituted... nothing.

In different ways, the critiques of Carson and Marcuse came to inform the view of science held by a generation of Western youth. When, back in the 1950s, the ex-communist critic Arthur Koestler had commented that reason 'begins with Galileo and ends with the hydrogen bomb'[41], few had listened. Now, as Theodore Roszak observed in his famous *The Making of a Counter Culture*, a radical rejection of science and technological values was 'close to the centre' of American society, rather than on the 'negligible margins'.[42]

But the logic of Marcuse was that the American bombing of Vietnam, for example, was an act born not of imperial *realpolitik*, but rather of the scientific mentality. Radical in form, it was a logic whose content was profoundly retrogressive. If, after all, the artefacts of humanity – laboratories, tools, technologies – were *inherently* flawed, then men and women would be unable to shape their own future. Furthermore, it followed from the nature of technology that Western élites could be no more responsible for the devastation inflicted by it than any other social group. Whatever Marcuse's leftism might lead him to say about them as people, if technology was the primordial guilty party, then those who took advantage of it were innocent.

In his the *Edge of Objectivity* (1990), Gillispie rightly notes the conservative character of the radical critique of science. For him the New Left

commingled a political with cultural hostility to exact science (although not, of course, to natural history, ecology, or organismic biology). In these respects, its position was identical with that of the Far Right in Weimar Germany.[43]

More recently, the widely-acclaimed critical writer on the history of Green movements, Anna Bramwell, compares the New Left with an even earlier generation of conservatives. Marcuse's central themes, she observes in her book *The Fading of the Greens*, 'parallel the cultural criticism of capitalism to be found in the works of nineteenth-century German and British conservatives.'[44]

Over science and technology, the New Left was in fact *more* backward-looking than its right-wing targets. As we have already seen, mainstream conservative and liberal thinkers could not be wholeheartedly against science; they were acutely aware that modern society could not function without it. But the New Left outlook was different. In opposing progress altogether, it castigated science without reservation.

That the New Left critique was inherently conservative allowed many of its ideas to be adapted by establishment opinion. Shorn of some of its more extreme postures, New Left prejudice against science and humanity's exploitation of its potential was given a ghostly but worldwide echo in the Club of Rome's portrait of doom to come, *The Limits to Growth* (1972).

The New Left's trajectory parallelled the integration of the Old Left into officialdom in the 1960s. By the 1970s, officialdom had itself abandoned the crusading zeal of the previous decade. And since that time, a preoccupation with natural limits and with the destructiveness of human activity has led to a further shift away from the themes of the Enlightenment. In the 1990s, the ideological premises of the New Left no longer represent a Counter Culture.

In their own ways, many of the nightmares of Carson and

Marcuse have come to form the basis of mainstream culture itself. The talk today is of a 'risk society', in which the risks are all manufactured by humanity. Risk is also, we are told, a phenomenon which is itself subject to that 'globalisation' which has transfixed every kind of commentator over the past decade. Altogether, we are told to fear a Revenge of Nature. Carson and Marcuse are, in this sense, alive and well and living in most parts of Western society.

We have come full circle. We saw how the right no longer has the confidence to uphold the market, on its own, as a solution to humanity's problems. Ironically, this has become clearer, the longer the West's triumph over the East has, since 1989, gone on. For the left, on the other hand, the collapse of the Soviet Union, together with the discrediting of Keynesian policies over the years, means that all its old alternatives to the market have been found wanting. Both right and left now find that the class conflict and mutual animosity which was put in aspic during the Cold War has been dissipated since its end. The convergence of the two sides today means that both accept the market, but without enthusiasm. What both really get worked up about, by contrast, is humanity's alleged Rape of Nature.

Obsessions after the Cold War (1): Ecology, Entropy and Limits

At the beginning of the twentieth century, as we have seen, the sense that the West had reached the end of the line led to a convergence between conservative and liberal thought about science. At the end of the same century, similar feelings have led to a merger of left and right views about science. The difference is that, today, political and social paralysis is more complete. There now seems no sense that a real renewal is possible. The end of the Cold War has unleashed a sense of foreboding which now affects every aspect of discourse on science.

Take the interminable discussion of low educational standards in general, and the problem of 'illiteracy' in science in particular. In his 1992 essay *The Advancement of Science*, Leon Lederman expressed high hopes in US Labour Secretary

Robert Reich. For Lederman, the 'top-priority' status that Reich had given to the re-educating of America was good news. But if, following Reich's bestseller *The Work of Nations*, both conservatives and liberals are now unanimous that education must be a priority, it is also evident that nobody knows what to do about it.

The view that a broad and scientific education can rescue society from its deep-seated economic and social problems speaks more of a culture of despair than of a rolling-up of intellectual and practical sleeves. It ought to be obvious that today's crisis in education is a *symptom*, not a cause, of those problems. By themselves, schools and universities are not magically going to mobilise science for good, whatever the plans now drawn up for them might suggest. Science education will only fully help humanity control its fate once other, more basic wellsprings of wealth and power – labour, capital, the state – are fundamentally transformed for the twenty-first century.

With science education we have but one example of stasis as a way of life. With the emotive issue of ecology, we have another. Nowadays, everyone agrees that the actions of humanity must and will be checked by *limits* which are given in nature itself. Humanity is widely seen to be in a stalemate of irrevocable constraints – a stalemate in which, ultimately, nature will cut the human race down to size.

In the late 1960s and early 1970s, pictures beamed back from the Apollo programme did much to convey the idea that human ambitions, despite the conquest of space, necessarily had to be finite. Now, more than 20 years later, the image of 'spaceship earth' has become so ingrained that there is little belief in the possibility of reversing the deterioration of the ecosphere.

Not long before he became US Vice President, Al Gore spoke of developing a 'Marshall Plan' for the environment. Now there is no such talk. Everyone now knows that little or no action has come out of the Rio 'Earth Summit' of 1992 on the global environment. Nobody, on either left or right, really expects anything to happen. Nature has limits, it appears; and men and women are fundamentally limited by their inability to undo the damage they have done to the planet over the past 50 years.

To explore this point further, we turn now to compare two books which are relevant to it: *The Future of the Market* by the left-wing German economist Elmar Altvater, and *Living Within Limits*, by the right-wing American ecologist Garrett Hardin.

Elmar Altvater and Garrett Hardin

As an incisive analyst of Germany's predicament, Elmar Altvater made his reputation in the late 1960s and early 1970s. Then, unlike many of his contemporaries, he felt that progress was possible. Today, however, Altvater has adopted the doctrine of limits with all the fervour of a late convert. He condemns human attempts to dominate nature in the language of Adorno and Horkheimer:

> Human emancipation is expected to arise through further domination of nature and through a transition from the domination of people over people to the rational adminis- tration of things – and yet humanity is itself a part of nature. The domination of nature and the administration of things lead to a new domination over people.[45]

This represents a considerable *volte face*. In the 1970s, Altvater wrote powerful critiques of society. In them he suggested that it was the accumulation of capital – even German capital – together with declining corporate profitability which impeded economic growth. Today, however, the butt of Altvater's polemics is not the slowness of capitalist society to expand wealth, but rather the presumption he believes right-wing economists show in their continued belief that a better life is within humanity's grasp. Referring to Milton Friedman's 'Chicago School' of partisans for market forces, Altvater writes:

> The neo-classical/neo-conservative 'counterrevolution' (Friedman) is an attempt not to stop the 'wheels of history', but to break free from historical inevitabilities, to

abstract from the historical, natural and thus social constraints upon accumulation.[46]

Here the right is upbraided for seeking too much of the future. For Altvater, human beings exhibit only self-delusion when they try to conquer history, nature, or themselves. The effort is futile. Our abilities are limited.

Garrett Hardin takes a similar line to Altvater. The major ways in which ecology and economics differ, he explains, is in their attitudes toward limits, toward 'discounting the future', and toward 'dealing with irreversible changes'.[47]

For Hardin as for Altvater, ecology must be superior to economics. Economics forgets that too much human intervention now can only lead to Final Retribution later.

Both Altvater and Hardin go on to generalise from this fatalism in a dramatic way. In a daring leap from the social sciences to natural science, they contend that modern economics, and in fact any human aspiration to progress, violates the second law of thermodynamics. This blunt claim deserves examination.

The second law of thermodynamics states that the useful forms of energy in any closed system decline over time; alternatively, that the disorder, or *entropy*, of any closed system increases over time. Altvater is very fond of this law. For him it means that humanity's ambitions to order nature can only lead to failure. Whatever the quantitative worth of artefacts in terms of *value*, their qualitative worth as objects of utility is subject to inexorable decay. Thus humanity may plan and implement, but its work will inevitably be unravelled... by nature. 'Economics', concludes Altvater, 'can no longer dispense with a theory of use-value'. In economics, indeed, the explicit concept of entropy should occupy 'a central place'.[48]

Hardin agrees. Approvingly, he quotes from a piece written by Daniel Underwood and Paul King in the first issue of *Ecological Economics*: 'The fact that there are no known exceptions to the laws of thermodynamics', these authors proclaim, 'should be incorporated into the axiomatic foundations of economics'.[49] This conclusion, Hardin strongly endorses.

What are we to make of Altvater, Hardin and entropy? Today, no reputable scientist believes the second law of thermodynamics to be wrong. It is often said that if a new theory is developed which contradicts the second law, it should be abandoned, no matter how good it looks. But the inviolability of a law to do with forms of energy says nothing about the future of social life.

The earth is not a closed system; it receives energy input from the Sun. Until humanity colonises space, our activities will be limited by the energy we receive from the Sun, supplemented by those useful forms of energy which are already stored up in the earth. However, this aggregated energy available to humanity is vastly greater than the amount currently in use. In this sense, there are no absolute limits to the energy at mankind's disposal, nor will there be for millennia.

Altvater and Hardin are aware of this fact. Nevertheless, they casually disregard it. Altvater consigns the real *limitlessness* of energy to parentheses. He prefers to highlight the saloon-bar wisdom that humanity is 'running out' of everything:

> The rate of entropy increase indicates that, in comparison with the *present* state of things, a lesser amount of useful energy and materials is available in the *future* (leaving aside energy input from another system, such as the sun).[50]

Altvater's literary-scientific sleight of hand shows that he is bent on setting limits to human ambitions, regardless of the facts. Altvater and Hardin also seem to be unaware that their daring intellectual breakthrough has bleak precedents. Spengler saw in the second law of thermodynamics both the *ultimate* limitation on human society, and the *immediate* cause of the onset of social decline. As Stephen Kern notes in his fascinating work, *The Culture Time and Space, 1880–1918*, Spengler argued: 'what the myth of *Götterdämmerung* signified of old, the irreligious form of it, the theory of Entropy, signifies today – *world's end as completion of an inwardly necessary evolution*'.[51]

From their bowdlerisation of the second law of thermody-namics, Altvater and Hardin move on to even more questionable terrain. They argue that all today's ills are the result of humanity pushing up against, and exceeding, natural limits of one sort or another. Altvater writes:

> The limits are there, even if social imagination is needed to visualize whether and in what way they are definitive. They express themselves in climate disaster, ecomigration, revolts against poverty and hunger, nuclear catastrophes or the danger of atomic war, and inbreeding due to narrower species diversity.[52]

Altvater and Hardin are, by today's standards, not at all eccentric. When they reel off every conceivable problem and proceed to lay the blame for all of them on humanity's disregard for natural limits, they merely express what has become the conventional wisdom. Together with a host of 'experts', the United Nations has turned this tendency into an art-form.

For the UN, social problems must always be situated in the larger, graver context of mankind over-reaching itself. In September 1994, the UN Conference on Population and Development managed to put the poverty of the Third World down not to debt or war, but to ignorant people breeding like rabbits. Humans have, it appears, asked too much of nature. Now, in a Faustian scenario, they will get their comeuppance – unless we (or, more accurately, people in the Third World) change our ways. By our irresponsible actions, we are adding to entropy and to chaos. In the face of an ecology of unalterable limits, we must learn a new humility.

Obsessions after the Cold War (2): Demography

Many of the themes coming out of the UN conference were anticipated by Paul Kennedy in his work *Preparing for the Twenty-First Century*. Kennedy wrote of trends which 'may even threaten the long-term existence of humankind itself'. The 'first and most important' of those trends, he argued, was 'the surge' in the earth's population and the rising demographic imbalances between rich and poor countries.[53] By the latter, he meant the fact that 97 per cent of today's population growth is in the Third World.

Projections of the size of the human population by the end of the twenty-first century give 10 or 11 billion as a likely figure. This, for Kennedy, would be a disaster:

> It is inconceivable that the earth can sustain a population of 10 billion people devouring resources at the rate enjoyed by richer societies today – or at even half that rate. Well before total world population reaches that level, irreparable damage to forests, water supplies, and animal and plant species will have occurred, and many environmental thresholds may have been breached.[54]

Kennedy thus held that the Club of Rome and others would be shown to have been right all along, even if their estimates of the precise timing of disaster were wrong.[55]

To his credit, Kennedy was at least honest enough to open his book with an explicit acknowledgement of his intellectual mentor – the Reverend Thomas Robert Malthus. Thus, in examining the way demography has joined ecology, entropy and limits as a post-Cold War obsession, we must begin our study with Malthus.

In his *An Essay on the Principle of Population as it Affects the Future Improvement of Society*, first published in 1798, Malthus argued that poverty was caused by an excess of people. Growth in population, he contended, progressed geometrically; that of the productivity of land, by merely arithmetic leaps and bounds. Thus, without 'checks' on its breeding, humanity lacked the capacity to feed itself.

At first, Malthus' views were well received. To the British ruling class, they provided a convenient account of how the hunger and unemployment brought on by the early stages of industrialisation were in fact the fault of the masses and their lust. But if support for Malthusianism has risen with each major recession since 1873, it has also fallen with each succeeding interval of prosperity. Malthus's arguments began to look ridiculous when populations rose, and economies still expanded.

Today, as food mountains attest, there is no natural reason why the Earth cannot support its human inhabitants. Half the land which could be farmed quite productively in the world is not yet farmed. As for soil degradation, it is not a consequence of over-use, but rather of economic policies and thus bad agricultural practices. In large parts of the Third World it is due to sheer economic plight.

Given the right economic conditions, African land could be farmed five times more intensively than it is. This level of intensive farming exists in America, yet America suffers far less degradation of its soil than does Africa. Again, and related to this, vast parts of the Third World are under-populated, not over-populated. It has been estimated that, *using agricultural techniques already in existence*, the Third World alone could feed 32 billion people, without the help of the vast fertile areas of Russia and the Ukraine.[56]

So why, then, do arguments based on natural limits now have such an appeal? Before *Preparing for the Twenty-First Century*, Kennedy's previous book, *The Rise and Fall of Great Powers* (1987), had concentrated on the decline of the United States. In his more recent work, however, he proved obsessed with the inability of every nation to shape global developments. After the end of the Cold War, his blues were deeply felt. Kennedy maintained that the advance of science and technology would only breach more natural limits, and thus cause more inequalities between rich and poor.[57]

This is the kind of standpoint which holds that God made a mistake when he let humanity discover fire. The only conclusion to be drawn from this is that people should curl up and hide from the world, in the hope of deliverance. For the moment, people still recoil from this conclusion. But the premises from which it follows have already become the dogmatic belief of millions.

The real economic limits and political fatigue which charac-
terise the West today are the main factors that lead to a mass
psychosis about limits in nature. In population matters, that
psychosis casts humanity as its own worst enemy. Facts can be
dispensed with. Instead of recognising the real and very
contemporary social obstacles to progress for what they are, the
'experts' revive an old English parson: Malthus. In so doing,
they make strictures against human procreation itself.

The Divorce of Science from Social Progress: its Consequences for Science

The cumulative effect of the trends we have discussed has
consequences not simply for society's perception of science, but
also for the development of science and scientists' perception of
themselves.

Everyone, notes Leon Lederman, gets depressed at times.
But what he hears inside the scientific community in America
brings about the very darkest of moods. There, depression is
'widespread, independent of institution, field, and rank'.[58] The
sources of this unhappiness are many and varied. The most
important are: the incoherence of government science policy;
the lack of a strategic vision for science; the shortage of cash for
scientific work of all kinds, but especially for fundamental
research; and the rush to make commercialisation the be-all and
end-all of scientific research. In his own report, Lederman
focuses on the last two.

Since 1968 and the end of the post-war 'golden years' for
science in America, US research funding has risen 20 per cent
in real terms. That sounds good; but as Lederman points out,
the myriad fields of genetics are but one example of how the
avenues of enquiry explored by science have multiplied rapidly
over the past 25 years. At the same time, the cost of experimental equipment has mushroomed. Just to keep pace with
these two changes, Lederman estimates, the level of funding
today would need to be double the 1968 level in real terms.

The most immediate problem facing American science is that young scientists are put off from pursuing research by poor career prospects. But perhaps more disturbing still are frictions within the US scientific community, and between it and representatives in other countries. Because rivalry is intense, scientists often fail to communicate with each other. The threat of usurpation by other researchers also leads to a cautious mentality. Together, a refusal to convey insights to others, and an unwillingness to strike out on new paths, act as a drag on truly innovative research. The result: science is further divorced from social progress.

Writing about the mail he has received from his peers all over America, Lederman comments:

> The letters reveal potentially important changes in the way scientists as individuals pursue their craft. As a consequence of the increasingly difficult search for funding, academic scientists are less willing to take chances on high-risk areas with potentially big payoffs. Instead, they prefer to play it safe, sticking to research in which an end-product is assured, or worse, working in fields that they believe are favoured by funding agency officials. These scientists are also increasingly viewing their fellows as competitors, rather than colleagues, leading to an increasingly corrosive atmosphere. The manifestations of this attitude range from a reluctance to share new results with other scientists to public bickering about relative priorities in funding different fields.[59]

Lederman has put his finger on real difficulties. Scientists involved in the Human Genome Project, for instance, are not as short of cash as those in other fields. Nevertheless, even they suffer from many of the same problems. Sir Walter Bodmer, Director of the Imperial Cancer Research Fund, reckons that many of the professionals in the Project spend more time and effort trying to patent their results than they do developing knowledge in the first place. On two or more occasions, international disputes about copyright have threatened to tear the whole endeavour apart.

Scientists do not just have to keep their eyes glued to what funding agency officials want, or to what other scientists are up to. More and more, they are forced to obey the commands of private companies. This commercialisation of science has led to myopia and the neglect of fundamental work. On top of this, however, there is another problem which Lederman rather understates: the overall social and political climate.

Cuts in funding hinder science as much through the demoralisation they engender as through the direct effects of not having enough money to do work. But declining morale stems also from sensitivity to a public opinion which is often indifferent, and sometimes even hostile, to science. Everyone works better when they think their work is appreciated; and the reverse also holds true. A simple example is Roger Gosden, the Edinburgh-based scientist who has done much of the pioneering work on the possible transfer of eggs from an aborted foetus to a mature female. His work could, in principle, be applied to the human species. This prospect caused an outcry in Britain in late 1993. Gosden has said he may abandon his current work. His reason: a desire to pursue a field of science which might be a little less controversial.

Lederman looks back on the 'golden years' of US science as a model for how it should be conducted – with plenty of money, and with plenty of moral support from government. His is a widespread, and not inaccurate, perception of the post-war science boom. Science *can* advance at a rapid rate, with a quite contented scientific community, even if there is little real sense of social optimism in society. Though it is always carried out in a given economic and political setting, the subject-matter of science is nature, something outside society. Accordingly, science has its own measures of success.

Yet the sustained economic expansion enjoyed by America from the 1940s to the 1960s is the exception, rather than the norm, in the twentieth century. Lederman's hope that America can recapture the scientific successes of this period, and his fear that Japan will soon repeat and improve upon them, seem misplaced. The past few years have shown that even Japan is by no means immune from a fragile world economy.

Even in the era for which Lederman is so nostalgic, the coherence of science policy was not so great. Two projects showed the potential for the planned application of scientific experience and know-how – the building of the atomic bomb during the war, and the Apollo missions to the moon. Not fidelity to progress, but war, Hot and Cold, was what made the American élite establish a dynamic behind science. It is an indictment of modern society that many scientists failed to find in peacetime an environment as conducive to work as that provided by the Manhattan project and a regime of intercontinental ballistic missiles.

The beginnings of economic downturn at the end of the 1960s very quickly led to a shift in élite attitudes toward the pursuit of organised science. The romance simply ended. Then, in turn, scientists were put on the back foot.

Like most other people, most scientists have lost faith in social progress. But this makes it hard for them publicly to uphold their business. Of course, the idea of social progress through greater human dominion over nature is different from the idea of science. Nevertheless, science cannot but be linked to such an idea. Science increases the possibilities for human action. Consequently, scientists are thrown on the defensive when they search for a reason to pursue science other than as a means to increase human mastery over nature. In the longer term, this can only have harmful consequences for the self-esteem of the scientific community and, ultimately, for the pursuit of scientific work itself.

Scientists don't always falter. About some things, they are very upbeat indeed. Still, many scientists can only maintain a sense of optimism today by *retreating into science*. Again, this tends to accentuate the divorce of science from progress.

Scientists rightly feel proud of what they have achieved this century. On the occasion of the 125th birthday of the leading science journal *Nature*, an editorial in that journal commented:

the past century has been a great adventure for all those concerned with science. It is hoped that the part of this issue labelled 'Frontiers of ignorance' will suggest how much excitement is still to come.[60]

'Frontiers of ignorance' itself began with the observation that 'the scientific enterprise has had a marvellous century'. Yet *Nature's* enthusiasm for the recent past and future of natural science forms a vivid contrast with many leading scientists' views on the willingness and ability of broader society to *act upon* the possibilities offered by science. That contrast is what we mean by scientists maintaining a sense of optimism only by *retreating into science*.

The retreat means that, when they come to communicate their enthusiasm for science to the outside world, scientists often stumble. A typical defensive response is to try to separate science from technology – praising the former, but conceding to criticism of the latter. Thus, in 1992, the *Bulletin of the Atomic Scientists* felt unable to celebrate the 50th anniversary of Enrico Fermi's achievement of the first controlled, self-sustaining nuclear chain reaction – because of the context in which this was done, and the application to which it was put. Similarly, the historian of science Timothy Ferris gives in to Bryan Appleyard's tirade, in *Understanding the Present*, inasmuch as he seeks only to re-direct it away from science and on to technology.[61]

The problem with this defensiveness is simple. If the idea is to save science by separating it from technology, scientists will be disappointed; for just as science has a clear relation to social progress, so it is bound to be linked to technology. Scientific advance necessarily suggests technological application. Reactions against technology are, moreover, inevitably reactions against science. Conversely, the general public will never have a durable passion for scientific knowledge in its own terms alone.

Lewis Wolpert, Chair of the Committee for the Public Understanding of Science, and John Maddox, editor of *Nature*, put the case for such passion with plenty of personal gusto.[62]

But the outcomes they search for are not forthcoming. Indeed, a one-sided defence of pure knowledge can backfire. By separating theory from practice, it calls into question not just technology, but the function and purpose of science.

Contemporary *anti-science trends* are bad news. But they are just a small component of a broader rejection of progress. Only a concerted drive to challenge the *anti-progress consensus* can offer a challenge to the suspicion with which science is greeted today. A failure to do this – and most if not all recent campaigns to popularise science do fail to do this – will leave anti-science trends untouched.

The Divorce of Science from Social Progress: its Consequences for Society

In their own Retreat from Reason, scientists make vain attempts to seek refuge in the pure. This is a loss not just for science, but also for society. The full potential of science is not realised. Mystical and religious ideas get a fuller rein. Also, as Lederman points out, scientific advance can inspire more people than just scientists.[63] Thus, even fine artists have something to lose when science slows up.

There is a more fundamental issue at stake here. When society abandons the goal of progress, expectations of what can be achieved using science are lowered. The reversible failures of society are viewed as the unalterable givens of nature; and that can lead to regrettable conclusions. Thus the 1990s had barely begun before an editorial in *The Lancet* backed a senior doctor's suggestion that babies suffering from diarrhoea in the Third World should be denied water, the better to alleviate overpopulation there.[64]

Whereas the immutability of nature is an unstated assumption in Kennedy's book, it is explicit in Paul Harrison's *The Third Revolution: Population, Environment and a Sustainable World*. Harrison runs through the enormous documentary evidence showing the gap between the promise and the reality of Third World agriculture. But he refuses to base his policy prescriptions on that promise. Instead he assumes that each country must feed itself, and only with existing levels of technology and skill.

This approach now dominates official discussion of the Third World. The free-marketeer's dogma that there is no such thing as a free lunch, the dismissal of progress through bogus applications of thermodynamics to society, the paranoia about population – these things reach their climax in the doctrine that, despite the existence of a longstanding division of labour in the world, each nation state must grow all its own food or be condemned to oblivion.

Harrison no doubt sees himself as a radical; but his narrow, localist framework makes low expectations a way of life. And, in turn, it must be said that this parochial perspective dovetails all too well with the right's growing tendency to offer low expectations about science as official policy.

Low Expectations as a Policy

Low expectations do not just come about because people in high places have reneged on the whole idea of progress. The lowering of popular expectations is a goal which élites actively strive for. Only by lowering expectations is it possible to distract the popular gaze from the economic and political barriers to progress which exist today.

Today, élites and their literary protagonists have felt the need to smudge over the contrast between the *promise* of science and its *limited practical effects*. Why is science 'not delivering' as far as billions of the Earth's inhabitants are concerned? Instead of looking at the priorities of social systems unwilling and unable to fund or realise science properly, many of those trying to answer this question have responded with a campaign to cut the pretensions of science down to size.

Bryan Appleyard, as we might expect, suggests that it is simply fruitless to enquire about humanity. Invoking Wittgenstein, he writes:

What we are is what we ordinarily are. This is what we do. We are our own embodiment. If you are possessed by the suspicion at this point that I have told you nothing that you did not know already, that is precisely my point.[65]

In her melancholic book *Science as Salvation* (1992), the philosopher Mary Midgley attacks those who would still celebrate science. Smugly, she declares that 'most of us have begun to see that the party is over. The planet is in deep trouble; we had better concentrate on bailing it out'.[66]

In this kind of sad prospectus, we find some distant feedback from the 'Two Cultures' debate which has set the tone for discussion about science in Britain since the early 1960s.

When C. P. Snow, the science mandarin, and, eventually, advisor to the Wilson government, first lamented the historic split, in Britain, between the 'two cultures' of natural science on the one hand, and the arts and humanities on the other, he struck a nerve. Snow's modernising plea for a higher status for science, and for more integration between arts and sciences, deeply disturbed cultural traditionalists. Years before Appleyard and others made a virtue of denigrating science and dampening down the human desire to be freer and happier, the influential Cambridge literary theorist F. R. Leavis replied in no uncertain terms to Snow's hypothesis that a vision for society and science could overcome personal distress:

> what is the 'social condition' that has nothing to do with the 'individual condition'? What is the 'social hope that transcends, cancels or makes indifferent the inescapable tragic condition of each individual'? Where, if not in individuals, is what is hoped for – a *non*-tragic condition, one supposes – to be located?[67]

For Leavis there was nothing but unremitting tragedy. Science could not alleviate that tragedy, and it was wrong for Snow to imply that it could.

The message we are given today is that scientists who look

forward to fundamental improvements in the human condition are deluding themselves when they are not just misleading others. The message is that it is always likely that there will be starving people in the Third World, for there are simply too many of them. We are told that it was merely naïve ever to have expected cheap electricity from atomic power. Likewise genetic engineering has, the *Economist* warns, 'over promised'; and the same magazine warns that humanity's urge to explore space only shows that 'dreams are the hardest thing to kill'.[68]

Earlier, we argued that sceptical attitudes toward science and hostile attitudes toward progress this century have been shaped by wars, recessions, and the now very evident defects of the market. The desire to lower expectations of science also springs from the gathering paralysis of Western economic and political life. Then, in the discussion of natural limits, both scepticism, hostility, and low expectations come into play at the same time.

A kind of downbeat narrowing of horizons, however, emerges most clearly in the philosophy of science. In Chapter One, we looked at how Popper and Kuhn expressed an anti-progress spirit in the years after the Manhattan project. Here, in looking at the philosophy of science of Pierre Duhem and Michael Polanyi, we show how lowering of expectations of science developed as a means of justifying distaste for progress.

Duhem and Polanyi: Lowering Expectations as Philosophy of Science

In contrast to Popper's focus on the conditional character of all knowledge, Duhem and Polanyi emphasised the immutability of certain truths. They defended metaphysics. Their stress on *apriori* belief was anathema to Popper; indeed Popper fell out with Polanyi, a contemporary, over precisely this point. Polanyi thought that unrestrained scepticism undermined faith and the stability of society, laying the basis for a totalitarian take-over. Popper's argument was the opposite. He felt that metaphysics was the *source* of totalitarianism, the only antidote being *doubt*. However, what both Polanyi and Duhem shared with Popper was the idea that human knowledge, and thus science, had finite boundaries.

Pierre Duhem was both a scientist and a philosopher of science in turn-of-the-century France. His ideas in both natural science and philosophy developed alongside those of Henri Poincaré.

In his studies of the transition from the medieval to the modern scientific worldview, Duhem stressed continuity over change. This made him a conservative in the true sense of the word. Duhem's conservatism was also evident in the way he constructed his theory of physical law. In *The Aim and Structure of Physical Theory*, first published in 1906, he made metaphysics separate from and impervious to the barbs of science: 'our system eliminates the alleged objections of physical science to spiritualist metaphysics and the Catholic faith'.[69] Indeed, only metaphysics could hope to explain the nature of physical reality: though 'no metaphysical system suffices in constructing a physical theory',[70] for Duhem, theoretical physics was 'subordinate to metaphysics.'[71]

Duhem did not stop there. Not only was physics subordinate to metaphysics, but metaphysics itself meant that the underlying nature of physical reality was beyond our ken: 'at the root of the explanations it claims to give', wrote Duhem, 'there always lies the unexplained'.[72]

In constructing his theory in the way that he did, Duhem established a seemingly rational limitation on rationality. Denying that the progress of science consisted in the attempt to grasp and then control physical reality, he restricted it, rather, to taxonomy. Duhem wanted to use powerful mathematical techniques to build that taxonomy, but expected and desired little more of scientific insights:

What is lasting and fruitful in these is the logical work through which they have succeeded in classifying a great number of laws by deducing them from a few principles: what is perishable and sterile is the labour undertaken to explain these principles in order to attach them to assumptions concerning the realities hiding underneath sensible appearances.[73]

For Duhem 'the realities' would forever remain unapproachable.

An accomplished chemist, the Hungarian Michael Polanyi was, through his post-1945 editorship of the journal *Science and Freedom*, later *Minerva*, a pivotal figure in twentieth-century philosophy of science. He agreed with many of the attitudes expressed by Duhem and, in particular, with Duhem's attack on Enlightenment thinkers for elevating reason above faith. Criticising Locke, and the Lockean emphasis on learning from experience, Polanyi wrote:

> We must now recognise belief once more as the source of all knowledge. Tacit assent and intellectual passions, the sharing of an idiom and of a cultural heritage, affiliation to a like-minded community: such are the influences which shape our vision of the nature of things on which we rely for our mastery of things.[74]

While, for Polanyi, there was an objective character to physical law, we cannot grasp it. All our knowledge contains an unfathomable element: it relies on what Polanyi, in the title of his major work, described as *Personal Knowledge*.

Like Duhem, Polanyi presented a seemingly rational restriction on rationality. The laws of nature, he argued, are such that humanity and its knowledge of nature are embedded in, and thus restricted by, the character of physical law. Influenced by one of Ilya Prigogine's early publications, Polanyi rejected the Enlightenment counterposition of humanity to nature. In its place, he favoured a teleological framework, in which human knowledge itself unfolded as a natural process – and could thus never anticipate it.[75]

Polanyi concludes his work with the view that the study of the teleological process in natural law is akin to 'how a Christian is placed when worshipping God'.[76] For Polanyi, science only confirmed humanity's diminutive stature compared with the natural world. Polanyi's religious undertones became overtones when he declared: 'unless you believe, you will not understand'.[77]

The stress on *apriori* truth made Duhem and Polanyi believe that metaphysics was the anterior and superior context for thinking about nature and natural science. Their philosophy could encompass the advance of science as technique and empirical knowledge, while ruling out the Enlightenment conception of social progress. In their prognosis, humanity could not aspire to take command of nature. Its knowledge of nature would always be limited. Anyway, humanity was merely a part of a larger religious or teleological process.

There are some similarities between this approach and that of Kuhn and the contemporary relativist theories of scientific knowledge. Both stress the limits on human knowledge which necessarily flow from the conceptual framework within which humanity studies nature. Indeed, Polanyi encouraged Kuhn's work, and Kuhn expressed his debt to Polanyi. To this day, it can be argued, relativism owes a debt to conservative thinking, and is often ready to allow religion to substitute for radicalism.

However, Polanyi differed from Kuhn. First, Polanyi was relatively bullish about science. As John Wettersten points out:

Polanyi emphasises the intellectual values of science and tackles these problems. The relativist reading of Kuhn, on the other hand, comes to the conclusion that sociology is enough. Polanyi sought to save the independent value of science and deemed scientific society something unique and of a high cultural value. The new sociologists explain all that away.[78]

Polanyi's explicit emphasis on a *fixed* metaphysics has obvious drawbacks – but it is at least honestly stated. Contemporary sociologists of science, in relativist mood, feel that their eclectic approach to science is better, because it prompts investigations of rival belief systems.[79] But this is not so. An accumulation of different subjective frameworks, none of which can be subject to Reason, is of dubious merit.

Polanyi's single framework for science is metaphysical; but it is also more optimistic, in that it at least tries to paint a big picture within which science as a whole plays an important part. Reflecting the more profound loss of certainty that has beset Western thinking about science over the past 30 years, any attempt to formulate such a total framework has been all but abandoned. As one recent survey of post-war thinking about science concludes, science has associations with 'the serviceable product of a home handicraft' rather than with 'an inspiring work of art'.[80]

Duhem and Polanyi, and also Popper, did not just suspend science from social progress as a result of their dissatisfaction with progress. They were also engaged in a *damage limitation exercise* – namely, one of explaining to modern society that progress in science need not mean improvement in humanity's lot. This apologetic intent is perhaps clearest in Duhem's preference for Catholicism over science; but it is also evident in Polanyi's work.

For Polanyi, inequality is unfathomable and unavoidable. Just as we must, intellectually, accept limitations on reason, so in practical terms we can do nothing about inequality. In *Personal Knowledge*, Polanyi wrote:

> The attempt made in this book to stabilize knowledge against scepticism, by including its hazardous character in the conditions of knowledge, may find its equivalent, then, in an allegiance to a manifestly imperfect society, based on the acknowledgement that our duty lies in the service of ideals which we cannot possibly achieve.[81]

Like Popper, then, Polanyi defended the status quo.

Ridiculing science when it looked too ambitious, Duhem and Polanyi sought to puncture the Enlightenment and expectations it had aroused. Their philosophy of science was fundamentally conservative. Science should know its place; no blame should attach to the social system which surrounded it. If science could not deliver, that was in the nature of science, not the current arrangements of society.

Anti-Science Trends Re-Considered – through the Example of Biology

Lowering expectations about what science can offer helps foster narrow horizons in society. But there are, as we have noted in our discussion of the climate since the end of the Cold War, other sources of conservatism. Today's fragmentation of politics and upheaval in economics adds to the conservative climate. Taken together, the different elements of conservatism make for suspicion of science.

Indeed, insofar as there are actually anti-science trends, as well as anti-progress trends in society today, it is because the advance of science raises awkward questions for a society fearful of change and of experimentation.

A reaction against science, because it poses awkward questions for Malthusian theories, lies behind Paul Kennedy's muted criticism of scientific advance. But what is muted in Kennedy is explicit in the pronouncements of others. *Sunday Times* columnist Norman Macrae writes in the *Economist* survey of *The World in 1995* about the problem of 'Too Many Oldies'. He can barely conceal his disgust with the achievements of modern medical science:

> We are keeping alive too many of the world's old people. Thanks to the computer, microsurgery and the revolution in genetic knowledge, we are going to advance in the next few decades to making nearly all diseases (including cancer, AIDS and heart trouble) rather easily curable.[82]

Macrae believes that medical marvels will lead to escalating demands on the state from a plurality of 'oldies'. He likes neither medicine, nor senior citizens.

Journalists like Macrae are made uncomfortable by scientific breakthroughs. So too are some doctors. For example, when a 59 year-old woman received treatment in Italy that allowed her to give birth, Dr John Marks, retired chair of the British Medical Association (BMA), insisted that the procedure bordered on 'the Frankenstein syndrome'. When a black woman in Britain gave birth to a white baby through egg donation, the BMA, pressed by the tabloid press for a 'line' on reproductive technologies and ethnicity, declared that 'every attempt should be made when employing these technologies to create as normal an outcome as possible'.[83]

The avoidance of artificial monsters, the insistence that things stay normal – increasingly, doctors today want to shout 'Enough!' and call a halt to medical intervention.

The problem for the high priests of British medicine, however, is that 'outcomes' from today's sciences are less and less 'normal'. Yet, as we noted in our Introduction, there is a burgeoning demand for fertility treatment in Britain today. The important issue is: what is medicine and science about if it is not about 'tampering with nature'? Why do august bodies like the BMA feel impelled always to raise the *ethical evils* of discoveries and experiments in genetics and the new reproductive technologies? Why do they not speak of the *advantages*, in terms of meeting human needs and aspirations, that the new methods promise?

With the help of the British government, the Human Fertilization and Embryology Authority (HFEA) conducted, in the first half of 1994, what was supposed to be an enquiry into the use of eggs from different sources in assisted human conception. Yet even before the enquiry had concluded its interview stage, members of the HFEA took it upon themselves to pontificate about the issue they were supposed to be investigating. Using eggs from aborted foetuses was, they argued, likely to cause public revulsion.

While officialdom more and more insists on reading intervention in the realm of human biology in the manner of Mary Shelley, even serious academic journals join in the general *hauteur* about science. Human gene therapy, for example, is described by a journal of the same name as the 'tyranny of the normal.'[84] Genetic screening is opposed because it will 'stigmatise people'. There are fears of clones, mutants, and the results that would follow the release of genetically engineered organisms. All in all, intervention into biology, especially human biology, is condemned today as never before – at just the point where its potential benefits are also unprecedented.

At the end of the twentieth century, science and genetics continue to advance, but stripped of most of their ties with social progress. Science is largely reduced to a theory of nature and a series of techniques. A system of ideas has been developed which can accommodate this dry diet of science as technique, but which actively *justifies* limitations on science as part of a wider justification for the lack of social progress. Gerald Holton, the eminent American historian of science, is therefore on the right track when he observes that

> to understand in more satisfactory terms what in fact is meant by anti-science, and what it may imply for the future of our culture, we must start with the recognition that no culture can be anti-scientific, in the sense of opposing the activity of 'science'.[85]

What Holton fails to contemplate is the idea that the origin of the contemporary dissatisfaction with science is to be found in the nature of modern society itself.

In his book *Science and Anti-Science*, Holton asks: 'could it be that, on having reached the end of the twentieth century, we will find that the widespread lack of a proper understanding of science itself might be either a source, or a tell-tale sign, of our culture's decline?'[86] Holton hopes that incomprehension of science is a source, rather than a sign, of the erosion of culture – for then science education could ride to the rescue. As for the origin of the problem, Holton pins much of the blame for contemporary anti-science trends on the cultural relativists.

But if lack of education were the problem, and if relativists alone were the root of anti-science trends, things would not have reached this pass. It is much more likely that anti-science trends indicate a broad *impasse* in society that no amount of inspiring science lessons can overcome. Indeed, the resurgence of anti-science trends around the 'final frontier' of biology speaks of that broad exhaustion of politics with which this chapter opened.

Science and Progress as a Problem   *Why?*

Modern society has failed to fulfill the promise of progress with which it was born. Twentieth-century thinkers have lost the faith in progress which previous generations took for granted. What is more, even promising progress can be problematic; failure to achieve what is held out as possible tends to rebound on those who make the promises.

During the Cold War, a Wilson could be strongly attracted to science and technology. So could a Kissinger; as late as the World Food Conference in Rome in 1974, the then US Secretary of State proclaimed:

> The profound comment of our era is that for the first time we have the technical capacity to free mankind from the scourge of hunger. Therefore today we must proclaim a bold objective: that within a decade, no child will go to bed hungry, that no family will fear for its next day's bread and that no human being's future and capacity will be stunted by malnutrition.[87]

But two decades have now passed since Kissinger's 'bold objective'; and we are no nearer reaching it. In the 1990s to hold out bold objectives is no longer fashionable, for delay in their achievement is incriminating.

Élites need science, but they are nervous about its associations with progress. Science is central to society, and its advance is an important goal for all governments. Yet it is profoundly alienated from the imagination of society. This is the real and abiding content of Snow's 'two cultures', and it is why debates on science will continue to be highly charged.

Chapter Seven:
The Loss of Certainty and the Quest for Beauty in Science

The Copenhagen interpretation of quantum mechanics, and the belief that chaos and complexity represent universal theories of nature, express a *Loss of Certainty* in both scientific advance and social progress. On the other hand, the *Quest for Beauty* in science – which elevates aesthetics over empirical evidence in the formulation of a theory – represents a retreat from material culture. Within the scientific community, both the *Loss* and the *Quest* are but symptoms of those broader, society-wide trends which we have already identified as *Science and the Retreat from Reason*.

The Forman Thesis: Quantum Physics as a Development Extrinsic to Science

In a very interesting and influential article, published in 1971 and titled Weimar Culture, Causality, and the Quantum Theory, 1918–1927, Paul Forman has argued that quantum theory emerged through the 'adaptation by German physicists and mathematicians to a hostile environment'. In the Weimar Republic, Forman contends, the ruling élite was racked by doubt and confusion about the future. Following defeat in the First World War, culture turned extremely hostile toward materialism, mechanism, and, above all, toward causality. In a neo-romantic era, German scientists developed an acausal quantum theory, Forman maintains, as a compromise with wider philosophical movements.

Elements of Forman's thesis were in fact first put forward by

211

the historian of science Max Jammer in 1966. Jammer maintained 'that certain philosophical ideas of the late nineteenth century not only prepared the intellectual climate for, but contributed decisively to, the formation of the new conceptions of the modern quantum theory'.[1] Jammer added that Wittgenstein's famous 1921 statement, 'Whereof one cannot speak, thereof one must be silent', had found 'an unexpected application in Heisenberg's new approach to the study of atomic phenomena'.[2]

Both Forman and Jammer are very persuasive. However, Forman goes too far in his 'externalist' account of the origins of quantum mechanics. It is worthwhile reviewing Forman's thesis here, because such an exercise can reveal much of what the loss of old, Newtonian certainties was all about.

In the aftermath of Germany's defeat in 1918, Forman's argument began, the dominant tendency in Weimar academia was an existentialist 'philosophy of life' which revelled in crises. Opinion swung against analytic rationality in general and against the exact sciences and their technical applications in particular. Indeed Forman asserts that just the word 'causality' symbolised, in those post-war years, 'all that was odious in the scientific enterprise'.[3]

Some of the scientists who were hostile to the intellectual trends of the time did comment on them. It struck Einstein, for instance, as 'peculiarly ironical that many people believe that in the theory of relativity one may find support for the anti-rationalistic tendency of our days'.[4] The German atomic theorist, Arnold Sommerfeld, responding to a request by a monthly periodical for an article on astrology, wrote:

> Doesn't it strike one as a monstrous anachronism that in the twentieth century a respected periodical sees itself compelled to solicit a discussion about astrology? That wide circles of the educated or half-educated public are attracted more by astrology than astronomy?... [We] are thus evidently confronted once again with a wave of irrationality and romanticism like that which a hundred

years ago spread over Europe as a reaction against the rationalism of the eighteenth century.[5]

Despite robust reactions like these, however, Einstein and Sommerfeld were exceptional. This by itself supports Forman's thesis that irrational moods had a wide influence.

Forman aimed to demonstrate that 'extrinsic' influences led physicists 'to ardently hope for, actively search for, and willingly embrace an acausal quantum mechanics'.[6] And this is indeed the conclusion which, at the end of his article, he came to. For Forman, the programme of *dispensing with causality in physics* was both 'advanced quite suddenly' after 1918 and, at the same time, achieved a very substantial following among German physicists well before it was justified by Heisenberg and Bohr. Forman also contended that

> the scientific context and content, the form and level of exposition, the social occasions and the chosen vehicles for publication of manifestos against causality, all point inescapably to the conclusion that substantive problems in atomic physics played only a secondary role in the genesis of this acausal persuasion, that the most important factor was the social-intellectual pressure upon the physicists as members of the German Academic community.[7]

So: while throwing the rationalist body out with the Newtonian bathwater was something an Einstein was not prepared to entertain, it was, as Forman indicates, something many, even most, German physicists did. They did this, Forman argues, because they were either directly influenced by the existing climate of opinion, or else because they felt compelled to alter the ideology and even the content of their science if they were to recover public approval after the débâcle of the First World War.

Forman's is a telling argument. He shows how Heisenberg and the main contributors to the Matrix formulation of quantum mechanics in 1925 nearly all hailed from upper middle-class academic families. The overwhelming majority were German; excluding their mentors, and their average age was 24 in 1925. This was just the generation that was most influenced by the spirit of the age.   Schrodinger 42

The Dane Bohr is the exception that proves Forman's rule. In 1925, Bohr was 39; but he had already adopted a philosophy at odds with materialism and mechanism well before the development of modern quantum theory. The historian Loren Graham recounts an anecdote among Bohr's Copenhagen associates to the effect that Bohr had already 'said everything' 12 years before quantum mechanics arose. The meaning behind this 'complimentary jest' as Graham put it,

> was that Bohr's philosophical disposition prepared him, even as a young man, for the passing of the mechanistic world picture, and that he saw the implications of the new physics more rapidly than did most of his colleagues.[8]

So far, so good. There can indeed be no doubt that the overall political and philosophical dynamics of those years did have a strong influence on the emergence of the quantum mechanical conception of the universe. In a second, we pass on to the still more important question of the *nature of the mediations* between the quantum and society. For the moment, however, it is notable that a number of critics of Forman have rightly replied that developments *intrinsic* to physics were the real midwife of quantum mechanics.

For such critics, it was the inability of classical physics to explain blackbody radiation, together with a whole range of further problems thrown up by the study of atomic phenomena, which led to the birth of modern quantum mechanics. In our opinion, Forman does exaggerate the importance of extrinsic factors. He argues:

> While it is undoubtedly true that the internal developments in atomic physics were important in precipitating this widespread sense of crisis among German-speaking

Central European Physicists, and that these internal developments were necessary to give the crisis a sharp focus, nonetheless it now seems evident to me that these internal developments were not in themselves sufficient conditions. The *possibility* of the crisis of the old quantum theory was, I think, dependent upon the physicists' own craving for crisis, arising from participation in, and adaptation to, the Weimar intellectual milieu.[9]

Forman is surely correct in identifying that an *acausal* quantum theory developed solely in response to the intellectual climate of the day. But he errs in believing that the crisis of the old theory has sources external to physics. The point is rather that the *acausal interpretation* of the new physics is the thing that, undeniably and entirely, emerged from the intellectual climate of the time. But *quantum mechanics as a mathematical formalism* arose in response to the crisis of the old (pre–1925) quantum theory.

To see extrinsic factors as clinching the necessary but 'not sufficient' internal factors behind quantum theory, as Forman does, seems to us too unmediated an account – despite Forman's best efforts. The real point about the intellectual climate upon which Forman rightly fastens is simple. The interpretation of the science could be taken to *be* the science, if it was assumed that quantum mechanics was a *complete, final* theory. Because many were predisposed, for wider social reasons, to seeing the Copenhagen interpretation as true, they drew the conclusion that quantum mechanics was a complete theory.

Accuracy about the real interplay of intrinsic and extrinsic factors surrounding the creation of quantum mechanics is important. It allows us to put into perspective an issue which Forman does not try to explain: why the Copenhagen interpretation of quantum mechanics established itself outside the Germanic world.

America illustrates how Copenhagen cast its spell. In America, quantum mechanics was imported wholesale from Europe. The Copenhagen interpretation was accepted uncritically, because it chimed well with a pragmatic urge to use the new formalism without thinking too hard about deeper philosophical questions. The fact that nearly all American scientists first met quantum mechanics at centres influenced by the Copenhagen interpretation was also critical. James Cushing sets the scene well:

> Prior to 1935, the major connection of young American quantum physicists (e.g., David Dennison, Robert Lindsay, John Slater, Harold Urey) was to the research centres in Copenhagen, Göttingen, and Cambridge (where the orthodoxy had already spread). Many American physicists were first alerted to the 'new quantum theory' by Max Born when he visited MIT and other U.S. institutions in the winter of 1925/26. Of thirty-two American visitors to Europe, all went to Copenhagen doctrine centres (and none, for example, to France). America was still in a learning, catch-up phase prior to 1930 and the pragmatic American approach to quantum mechanics led to its acceptance there, by and large without philosophical qualms.[10]

Such factors can explain the early dominance of the Copenhagen doctrine. But to account for its continued hegemony, we must look further. It can only be this century's loss of faith in progress which has given Copenhagen the worldwide sweep it has today. Copenhagen has become the conventional wisdom for both scientists and interested non-scientists alike. With its pessimistic emphasis on the limits to human knowledge and control over nature, the interpretation forms an essential part of the contemporary *Zeitgeist*.

Overall, the Forman thesis may allot too great a role to extrinsic factors in the infancy of quantum mechanics. But such factors have a weight which cannot be underestimated if the global influence of the Copenhagen interpretation is to be comprehended in all its persistence.

How Complexity and Chaos Come to Denigrate Humanity

Because the scope of their claims is greater, those who like to generalise from chaos and complexity theories often reach conclusions which are even graver than those drawn by proponents of the Copenhagen interpretation. Not only all of nature, but the whole of human society is said to be governed by laws beyond our control.

The grandiose scope now attributed to chaos and complexity shows that a kind of empire-building is at work with these theories.[11] Advocates of chaos and complexity want to apply them everywhere. Their promiscuous approach follows from three premises: that the development of all systems depends so intimately on their condition at any one time as to render prediction and control impossible; that order in society is intimately dependent on order in nature; and that order in nature is finely balanced, and liable to be upset by human intervention.

But where do these premises come from? The problem with the schemas put forward by the sages of complexity and chaos is that their ends so often justify the 'theoretical' means. In practice this ensures that it is the *results* of enquiries into complexity and chaos which frequently make an appearance as the *premises* of those enquiries.

That, it does not need underlining, is the antithesis of the scientific method. Yet, in the writings on the subject by leading players in complexity and chaos, the *desire* to establish the universal claims of the theories is overbearing. Preference takes the place of precision. Thus, Murray Gell-Mann attests himself deeply committed to the view that humanity and nature together constitute one big complex adaptive system. He presents this idea as the conclusion of his study of complexity. However, the chapter he devotes to naturalising the human condition lacks the rigour of his earlier chapters on the *mathematics* of complexity.

Stuart Kauffman is a similar case. He acknowledges that he has little empirical evidence to show that nature obeys the mathematical models he has developed. Nevertheless he is 'absolutely in love' with his central hypothesis. Again, emotion tends to triumph over judgement.[12]

In the book *Artificial Life: The Quest for a New Creation*, Steven Levy interviews many of the leading theorists in complexity. It is clear from his account that the misanthropic conclusions that many of his interviewees draw from their work are already firmly held, and in fact inform the whole 'quest' for the new 'creation' of which Levy writes. Chris Langton, at the Santa Fe Institute, is one of the leaders of the field. This is how Levy sums up his goals:

> Human beings will see themselves in a different light. We will not be standing at the pinnacle of some self-defined evolutionary hierarchy, but will rank as particularly complex representatives of one subset of life among many possible alternatives.[13]

The Green desire always to take humanity down a peg or two from 'the pinnacle of some self-defined evolutionary hierarchy' is one of the driving forces of excess in complexity theory.

What needs to be *demonstrated* by the theory is, rather, *assumed*. Human beings are nothing special – they are just 'one subset' of life. Once this humble proposition is accepted, all efforts are made to show that men and women should not get ideas above their station. Thus does complexity denigrate humanity.

Enthusiasts for chaos theory are driven by a similar sense of fatalism. They have set their face against the efficacy of human action, and that is all there is to it. We cannot change the big things; all we can expect is to create some small islands of order out of a sea of chaos.

Those who uncritically uphold complexity and chaos have tried to bolster their case by situating the theories in a historical and sociological context. The typical assumption made is that the universal character of these theories has been ignored until now because it would not have found favour in cultures committed to dominating and controlling nature. Stephen H. Kellert explains the 'nontreatment' of chaos by claiming that 'a social interest in the *quantitative* prediction and *dominating* control of natural phenomena contributed to the neglect of the study of chaotic behaviour'.[14] He goes on to argue that 'the mechanistic view of the world served as a legitimating ideology for the project of dominating nature, while at the same time functioning to secure a hierarchical social order'.[15]

For ourselves, we believe that it is not complacency or neglect of complexity and chaos which deserve attention, but rather the speed and unquestioning manner with which these theories have been adopted. With some but not all systems, theories of chaos and complexity undoubtedly offer insights. Many problems addressed by scientists in the field have, after all, refused to yield to traditional approaches. But then something untoward occurs. Frustration tempts scientists into speculating that complexity and chaos will provide a suitable model. Indeed, frustration is now so great that the leap to give these theories a universal application is easily performed. After that, sure enough, a Great Levelling takes place, in which the pretensions both of science (to uncover nature) and of society (to push ahead) are dismissed as utopian dreams.

Loss of certainty in progress reigns supreme. It is not only the outcome of complexity and chaos theories, but their unconscious inspiration.

Over the Top against a Theory of Everything

Chaos and complexity theories, today, are more and more seen as harbingers of a new paradigm in thinking about natural law. In this they are often pitched against a competing paradigm, popular in particle physics and in cosmology: that which sees a search for a 'Theory of Everything' as worthwhile.

In 1992, the journal *Science* published a summary of research in theoretical physics entitled 'The Quest for a Theory of Everything Hits Some Snags'. It began:

In the 1980s many leading physicists thought they had caught a glimpse of their finish-line – they believed they were closing in fast on a final, all-encompassing theory that would serve as a fundamental framework for physics. They even went so far as to talk of a 'theory of everything.' But now, like marathoners who have 'hit the wall', they are sagging under the weight of the effort – in this case the intractable mathematics – and are wondering whether the finish line is in fact an illusion.[16]

A *Scientific American* survey later on in 1992 raised more fundamental doubts. It presented two alternative scenarios. Science could be approaching a final theory. On the other hand, science might be breaking up into a disparate collection of areas of knowledge:

A slew of recent books by physicists – bearing such titles as *Dreams of a Final Theory* and *The Mind of God* – even suggest that physics might be approaching a fundamental description of matter and energy, sometimes called a 'theory of everything'.

At the same time, philosophers are insisting that, far from converging on the truth, science is degenerating into an ever more esoteric and fractious enterprise that offers no coherent vision of reality.[17]

The *Scientific American* writers were deeply sceptical of a final theory. Recent studies of so-called chaotic and complex phenomena – 'from gushing faucets to stock markets' – did, in their opinion, 'strike a blow against the facile reductionism underlying many predictions of completion'.[18]

But if anyone is guilty of 'reductionism', it is those partisans of complexity theory who use it to reduce human drives and desires to the level of a natural system. So it should be no surprise that, when the astronomer John Barrow draws upon chaos and complexity theories to raise justifiable doubts about the possibility of a Theory of Everything, he also begins, in an unjustifiable way, to write-off the possibility of fundamental knowledge altogether.

As Barrow sees it, the emphasis on *general behaviour* and *invariance* which characterises Theories of Everything should give way to a focus on the particular. In the past decade, he argues,

> there has grown up a renewed interest in the particular rather than the general. This, as we have seen in our earlier discussion of symmetry-breaking, has been brought about by a recognition of the extraordinary richness displayed by the outcomes of laws of nature that is not showed by the laws themselves. This study of outcomes has focused upon the evolution of complex systems, symmetry breaking, and chaotic behaviour. In all these things, time is of the essence. Invariance plays a weak role that sheds little or no light on the essential properties of the phenomena in question.[19]

The growing consensus, illustrated by Barrow, is that the Theory of Everything paradigm is more in tune with the thinking of the Enlightenment and progress, while the chaos/complexity paradigm is more in tune with the uncertain mood of today.

As a result of this, attacks on a Theory of Everything have a pronounced ferocity about them. That is why, for example, Appleyard chooses to single out Stephen Hawking for especial condemnation:

> Hawking has earned his £4m-plus by becoming a magician, a prophet, God's interpreter. He has also earned it by exemplifying and endorsing the current triumphalism of science, its ability to say anything it likes about anything it likes from behind a barrage of absent equations.[20]

The grain of truth in this is that, by contrast with the loss of certainty peculiar to chaos and complexity buffs, those scientists engaged in the search for a Theory of Everything do express a confidence in the powers of scientific thinking. However, thinking in the Theory of Everything paradigm is as much influenced by the apprehensive atmosphere of the 1990s as are thoughts inspired by universal applications of chaos and complexity theories. In particular, the search for a Theory of Everything can easily become a quasi-mystical enterprise which we label the *Quest for Beauty*.

Roger Penrose, Superstrings and the Limits to Speculation

Writing of his own work in quantum gravity in *A Brief History of Time*, Hawking comments: 'I'd like to emphasise that this... is just a proposal. It cannot be deduced from some other principle'.[21] He then proceeds to say:

> Like any other scientific theory, it may originally be put forward for aesthetic or metaphysical reasons, but the real test is whether it makes predictions that agree with observations. This, however, is difficult to determine in the case of quantum gravity.[22]

The problem Hawking points to would be instantly recognisable to any theoretical physicist. Much of the new work in the field is very speculative, and impossible to verify through experimentation. What Hawking points up, though, is the need to carry out an assessment of the role that speculation plays in modern science.

Roger Penrose, Rouse Ball Professor of Mathematics at Oxford University and a close colleague of Hawking, has done just that. He breaks down theories into three broad categories: *superb*, *useful*, and *tentative*. Relativity, quantum mechanics and Newtonian mechanics are superb theories. For a theory to qualify as superb, Penrose does not deem it necessary that it 'should apply without refutation to the phenomena of the world', but rather that the range and accuracy with which it applies 'should, in some appropriate sense, be *phenomenal*'.[23]

According to Penrose, only the quark model of subatomic

particles, together with the standard models of particle physics and of 'Big Bang' cosmology, firmly belong in the *useful* pigeonhole. There is, he says, some good experimental evidence for these theories. However, their observed accuracy and predictive power, at present, fall some way short of the 'phenomenal' standard required for them to be regarded as *superb*.

Penrose classifies a theory as *tentative* if it has no empirical support of any significance. Into this class Penrose places much of modern theoretical physics – including, for example, not just quantum gravity, but also superstring theory.

Penrose believes that some tentative ideas may 'contain the seeds of a new substantial advance in understanding'. At the same time, others are 'definitely misguided or contrived'. Finally, Penrose resists the temptation to split off a fourth category – *misguided* – from the 'respectable' *tentative* only by the thought of losing half his friends.[24]

The *problem* of the untestability of modern theory is felt so acutely precisely because the issues being studied by speculative approaches lie at the heart of theoretical physics. A Theory of Everything – one that would unite all the known forces and particles in nature into a coherent whole – would be a powerful thing. However superstring theory, the latest candidate for a unified description of all the forces and matter, is very far removed from experimental testing.

Superstring theory assumes that all observed particles are different manifestations of the same fundamental entity. According to the superstring idea, all particles which were previously thought of as little points are in fact not points at all, but little loops of 'string' which move through space, oscillating. Superstrings are conceived as objects which vibrate in 10 dimensions. The different levels of vibration correspond to the different particles observed in subatomic physics.

The theory also predicts, however, that the postulated deep-down merging of particle identities can only be observed under conditions of heat and pressure that cannot be produced by technology – now, or in the foreseeable future. There is thus less evidence for string theory today than there was for general relativity in 1915.

One of the attractions of superstring theory is that it combines general relativity with quantum mechanics. There is no law of nature that requires these two theories to fit into a single model. But the impulse for unity and simplicity has, over the years, grown so strong that theorists have pursued a quantum theory of gravity, without success, for decades.

The proponents of superstring theory justify their creation by pointing to its elegance, coherence and beauty. Amidst the excitement and enthusiasm which greeted early work in the field, Michael Green, one of the founders, argued: 'the theory's so beautiful it must be true'.[25] But since then, Green has back-pedalled a little. The idea of superstrings has, he says, 'generated too much hype'.

That is an understatement. Many still regard and promote superstrings as the foundation of a Theory of Everything; but Green himself now admits that 'it's an unfinished story. Superstrings are an approximation to something, and we don't yet know what it is'.[26] Leading string theorist Ed Witten offers an explanation of why that ignorance continues:

> It's been said that string theory is part of the physics of the twentieth-first century that fell by chance into the twentieth century... What should have happened by rights, is that the correct mathematical structures should have been developed in the twenty-first or twenty-second century, and then finally physicists should have invented string theory as a physical theory that is made possible by those structures.[27]

The problem as experienced by theoretical physicists, then, is that much of contemporary theory is *untestable*. The response of many to this situation is, however, misguided.

The sensible conclusion to draw from the current state of affairs in science is to recognise that much of what is called theoretical physics is in fact interesting mathematics, but not really physics. Yet this is not the conclusion drawn. Worse still, as Green's remarks indicate, some physicists are inclined to make a *virtue* of their heavy reliance on mathematical consistency as opposed to physical information.

The search for mathematical consistency can and has led to an idealist and Platonic view of science. The search feeds off the gap between theory and experiment that exists today; but it is not caused by this gap. Rather, it expresses a broad retreat from material culture, and a feint in the direction of religion.

When Speculation Masquerades as Theory

There is nothing to object to in speculation which is honest in its intent. Problems only begin when speculation is not recognised as such: in other words, when it is presented as an explanation of the physical world. For example, Penrose appears not to heed his own warnings. His conclusions concerning human intelligence are just as speculative, if not more so, than superstring theory. This is clear in his discussion of quantum mechanics in human mental processes:

> I am speculating that the action of conscious thinking is very much tied up with the resolving of alternatives that were previously in linear superposition. This is all concerned with the unknown physics that governs the borderline between **U** and **R** and which, I am claiming, depends upon a yet to be discovered theory of quantum gravity.[28]

Penrose hopes that an answer to the famous 'Mind-Body' conundrum, which has baffled scientists for centuries, may be found. But he wants to find it in a theory that physicists have not yet formulated – and one which we know from Hawking will be pure speculation if and when they do!

Snags begin to multiply when mathematical beauty is taken as a guide to physical research. Over the course of the twentieth century, the *Quest for Beauty* has become the guiding spirit in the leading branches of theoretical physics. The Nobel Prize winner Paul Dirac was an early twentieth-century exponent of views which have achieved a much wider resonance today. He wrote:

> It is more important to have beauty in one's equations than to have them fit experiment... It seems that if one is working from the point of view of getting beauty in one's equations, and if one really has a profound insight, one is on a sure line of progress.[29]

So: the balletic poise of an equation counted for more that it fitting with experiment.

In fact, Dirac's whole bent was to downgrade the significance of experimental results. Toward the end of his life, in 1978, he gave a paper, 'The Excellence of Einstein's Theory of Gravitation', to a UNESCO symposium on the impact of science on society. After running through the various experimental verifications of Einstein's ideas on gravitation, he proceeded to deny their force:

> The Einstein theory of gravitation has a character of excellence of its own. Anyone who appreciates the fundamental harmony connecting the way Nature runs and general mathematical principles must feel that a theory with the beauty and elegance of Einstein's theory *has* to be substantially correct. If a discrepancy should appear in some application of the theory, it must be caused by some secondary feature relating to this application which has not been adequately taken into account, and not by a failure of the general principles of the theory. One has a great confidence in the theory arising from its great beauty, quite independent of its detailed successes. It must have been such confidence in the essential beauty of the mathematical description of Nature which inspired Einstein in his quest for a theory of gravitation.[30]

Dirac's cavalier attitude to 'discrepancy' and to 'detailed

successes, was of a piece with his contemptuous posture toward the need for equations to fit experiment.

Hermann Weyl, a mathematician specialising in relativity theory, gave Dirac's approach even more forthright expression: 'My work', he wrote, 'always tried to unite the true with the beautiful'. But when Weyl had to choose one or the other, he reported, 'I usually chose the beautiful'.[31]

Weyl was a contemporary of Dirac's. Together, and with many others, they made sure that the *Quest for Beauty* in science gained a vital and early foothold in the twentieth century. Of course, the *Quest for Beauty* had a long pre-history of enthusiasts. However, in the Renaissance and the Enlightenment, the desire to aestheticise the world was kept in check by the prevailing strength of experimental philosophy. It is this check that has been removed today.

Lessons from Kepler

The astronomer Johannes Kepler [1571–1630] developed a model of the solar system as the most beautiful geometric construction which he could envisage. For Kepler, mathematics was 'the archetype of the beautiful'.[32] He believed that he had successfully shown that each of the planets in the solar system orbited on spheres, each of which in turn nested within and without the five regular-sided three-dimensional structures which geometers had known since Greek times.

Kepler regarded the delicate proportions of his structure as his masterpiece. He undoubtedly had a creative imagination; he also had great mathematical ability. But Kepler also suffered from poor eyesight, and, for direct observation of the planets, only had access to a modest telescope. His good fortune was to work subsequently in Prague with the great, painstaking and rather reserved Danish astronomer, Tycho Brahe [1546–1601].

Brahe used his better vision, together with superior telescopes, in a lifetime study of the stars and planets. After his death, Kepler was forced to abandon his neat, nesting model of the solar system in favour of more prosaic laws – laws which, however, accurately reflected the elliptical orbits of the planets around the sun. In the age of the scientific revolution, then, empirical study and experimentation, not metaphysical speculation, were the key to theoretical advance. Despite his own annoyance with Brahe's caution about any hasty use of his observational data, Kepler could not but acknowledge the pivotal contribution that Brahe had made:

> Since divine goodness has granted us the most diligent observer, Tycho Brahe, from whose observations the error in this calculation of eight minutes in Mars is revealed, it is fitting that we recognise and make use of this good gift of God with a grateful mind.[33]

We can see here how the nonchalant attitude to the truth and to 'discrepancy', which we saw exhibited by Dirac, had worthier precedents with Kepler and Brahe. Before they intervened, a mathematical model which made an error of less than 0.08 per cent compared with the observed position of Mars would have been celebrated as a major achievement. But in the age of the scientific revolution, the new context of advanced empirical observations insisted that such a model be abandoned. Instead, a model was developed which sacrificed formal beauty for accuracy.

In cases where there was no possibility of testing out a physical model, scientists after Kepler tended to avoid metaphysical speculation. Such was their belief that experimental data should decide matters, even if that data was lacking. For example, although he was a great believer in an ordered cosmos designed by God, Newton refused to let this belief influence his theory. He wrote in his *Principia*:

> Hitherto I have not been able to discover the causes of those properties of gravity from phenomena, and I feign no hypotheses; for whatever is not deduced from the phenomena is to be called an hypothesis; and hypotheses, whether metaphysical or physical, whether of occult

qualities or mechanical, have no place in experimental philosophy.[34]

Since Newton's day, however, 'hypotheses' in the head have, in terms of their influence on the world of physics, acquired a life of their own. It has become acceptable to justify a theory purely on aesthetic grounds, irrespective of empirical evidence to back it up.

Because the advance of science in the twentieth century has been accompanied by a growing tightness of budgets and a sense of *impasse* in politics, a gap has opened up, as we have already discussed, between scientific theory and practice. The gap was not too noticeable in the first three decades of this century: both quantum mechanics and relativity theory were developed in a manner which allotted a major role to experimental confirmation.[35] But since the Second World War, at least, what was merely a gap has grown into a yawning chasm.

So little store is now set by practice, and so much by speculation, that Queen Beauty has made a come-back to outshine her reign in the time of Copernicus and Kepler. However, the contemporary distance between theory and practice is not the *cause* of the *Quest for Beauty*. That must be sought in the impact of broader trends in society and politics upon the thinking of physicists. *The Quest for Beauty* in the twentieth century is but a further example, in the field of theoretical thinking, of the trend we have termed the *Retreat from Reason*.

The Flight into Mathematics

The deliberate and often irrational aestheticisation of the natural world is expressed, firstly, in the mathematicisation of nature. This is how Morris Kline describes modern developments:

> Whereas mathematics served previously to represent, study, and advance the mechanical analysis of phenomena, today the mathematical account is fundamental. In fact, the mechanical one has been abandoned except perhaps in

very limited areas. The essence of any modern physical theory is a body of mathematical equations. Thus differential equations which in Newton's day were the servant of physical thought have now become the master.[36]

This development is, it is true, partly an unavoidable consequence of the separation of theory and practice. But it also contains an element of escapism: escape from the material world, from action and from change. Bertrand Russell brings out the point clearly.

Russell's explanation of his personal interest in mathematics is a telling one. He wrote:

> Remote from human passions, remote even from the pitiful facts of nature, the generations have gradually created an ordered cosmos, where pure thought can dwell as in its natural home and where one, at least, of our nobler impulses can escape from the dreary exile of the actual world.[37]

In *form*, Russell's own Retreat from Reason appeared *rational*. The desire was to make contact with an 'ordered cosmos' of pure truth.

Russell's path was by no means a lonely one. He joined with the philosopher G. E. Moore and the economist John Maynard Keynes in attacking the idealistic doctrine that reality is a human invention or creation. The three luminaries preferred to believe that the aim of knowledge was to make contact with a structured realm outside of human consciousness. But in perceiving this realm to consist of aesthetic, ethical and logical truths, their schemes were, as Robert Skidelsky remarks, 'akin to Platonic Ideas'.[38]

The *content* of the Big Three's ideas was thus nowhere near as rational as their form. Their thrust was away from engagement with the natural and social environment and toward mathematics, logic and aesthetics. Nor were British thinkers alone in this mathematicisation of the world. Erwin

Schrödinger, influenced by his experiences in the Austrian army in the First World War, also recommended resort to an overwhelmingly mathematical universe.

Schrödinger read the complete works of Schopenhauer in 1918. He also studied Spinoza intensely. The influence of the latter can be seen in Schrödinger's search, throughout his career, for a final system of mathematical truths that would capture the nature of physical reality. This stable mathematical world, Schrödinger contrasted favourably to the changing world of man. In 1940, he wrote:

> There is no worldly truth but mathematical truth. In politics, diplomacy, history, truth changes from day to day, and people get different concepts right. But mathematics never lie.[39]

This emphasis on mathematical truth led Schrödinger in the direction of a Quest for Beauty. Paul Dirac wrote:

> Of all the physicists I met, I think Schrödinger was the one that I felt to be most closely similar to myself. I found myself getting into an agreement with Schrödinger more readily that with anyone else. I believe the reason for this is that Schrödinger and I both had a very strong appreciation of mathematical beauty and this dominated all our work. It was a sort of act of faith with us that any equations which describe the fundamental laws of nature must have great mathematical beauty in them. It was like a religion with us. It was a very profitable religion to hold and can be considered as the basis of much of our success.[40]

Here we need only note the concluding nod in the direction of religion. From beauty in and of itself, it is a short step to Magic and then to God. At all events, an 'act of faith', as Dirac puts it, becomes a substitute for useful insights into the physical world.

Empirical Results Temporarily Postpone Operations

Schrödinger and Dirac held as strongly to a notion of beauty as any modern theorist. However, the high quantity and quality of empirical results obtained during their time prevented science from joining even the Nobel Prize duo in an all-out hunt for beautiful theories. General relativity had to test itself in response to clearly defined experimental and observational problems: the perihelion (drift) of the orbit of the planet Mercury; and the bending of light from a distant star by the gravitational effects of the Sun. After 1925, too, quantum mechanics established itself through its ability to resolve a broad range of disparities between theory and observation in the sub-atomic realm.

Here, as ever, the *interpretation* put on a theory was of course shaped by the broader spirit of the times. However, empirical results limit interpretations: these, no matter how beautiful they may be, will gain few adherents if they are contradicted by a substantial body of empirical results. At the very least, Forman can be said to have shown that the interpretation given quantum mechanics was influenced by the spirit of the times. The same social pressures that helped establish an acausal interpretation of quantum mechanics also pulled theoretical scientists in the direction of a Quest for Beauty. A complete lurch toward the latter trend was kept in check, however, by the strength of empirical results.

The interplay of the different trends, cultural and empirical, can be seen clearly in Heisenberg's reflections on beauty. Heisenberg was attracted by Platonic ideas. Take, for example, this personal recollection:

In the spring of 1919, Munich was in a state of utter confusion. On the streets people were shooting at one another, and no one could tell precisely who the contestants were... Pillage and robbery (I was burgled myself) caused the term 'Soviet Republic' to become a synonym of lawlessness, and when, at long last, a new Bavarian government was formed outside Munich, and sent its troops into the city, we were all hoping for a speedy return to more orderly conditions... Quite often it happened that, after spending the whole night on guard in the telephone exchange, I was free for a day, and in order to catch up with my neglected school work I would retire to the roof of the training college with a Greek school edition of Plato's Dialogues. There, lying in the wide gutter, and warmed by the rays of the early morning sun, I could pursue my studies in peace, and from time to time watch the quickening life in the Ludwigstrasse below.[41]

Heisenberg became increasingly attracted to Platonic ideas as he grew older. However, empirical results, and more importantly a belief that empirical results mattered, limited his reliance on beauty as a guide to research.

In 1926, Heisenberg gave a lecture at Berlin University on quantum mechanics. Among the audience was Einstein. He invited Heisenberg to walk home with him, so that they could further discuss the new ideas Heisenberg had presented. During their walk, Einstein asked: 'How can you really have so much faith in your theory when so many crucial problems remain completely unsolved?'. Heisenberg replied: 'if nature leads us to mathematical forms of great simplicity and beauty... that no one has previously encountered, we cannot help thinking that they are "true", that they reveal a genuine feature of nature'.[42]

To his credit, however, Heisenberg was, in general, more cautious than Paul Dirac, and certainly showed more *sang froid* than Hermann Weyl:

You may object [said Heisenberg] that by speaking of simplicity and beauty I am introducing aesthetic criteria of truth, and I frankly admit that I am strongly attracted by the simplicity and beauty of the mathematical schemes with which nature presents us... the simplicity of the mathematical scheme has the further consequence that it ought to be possible to think up many experiments whose results can be predicted from the theory. And if the actual experiments should bear out the predictions, there is little doubt but that the theory reflects nature accurately in this particular realm.[43]

The weight Heisenberg was prepared to put on 'actual experiments' to 'bear out the predictions' is clear enough.

With Heisenberg, experiment limited an overly-aesthetic conception of the universe. But indicative of the cultural pressures of the time is the fact that scientists of Heisenberg's generation who were not so constrained by, or concerned with, experimental validation engaged in the *Quest for Beauty* with as much fervour as Dirac. And as with Dirac, the quest smacked of religion. A mourner at the funeral of G. H. Hardy, Britain's leading pure mathematician in the early part of this century, said that although Hardy was an atheist, he had a

profound conviction that the truths of mathematics described a bright and clear universe, exquisite and beautiful in its structure, in comparison with which the physical world was turbid and confused. It was this which made his friends... think that in his attitude to mathematics there was something which, being essentially spiritual, was near to religion.[44]

In short, atheism in religious affairs was, for Hardy, no barrier to spiritualism in physical ones.

The similarity between the spiritual impulse and the thinking behind much of theoretical physics today might suggest that, although science still competes with religion, it now does so only as a variety of the same species. One scientist who would dispute this is Steven Weinberg. A Nobel Prize winner in physics, Weinberg is an unrepentant atheist and has little time for philosophers. Yet though he is well aware of the spiritual dimension to the search for beauty, he firmly believes that the *Quest for Beauty* in science is a perfectly rational enterprise.

Steven Weinberg, Simplicity and Inevitability

At a meeting of Britain's Royal Society some years ago, Weinberg focused on the cleavage between theory and experiment. Quantum gravity, he said, was 'inaccessible to any experiment we can devise'. He then went further, arguing that 'physics in general is moving into an era where the fundamental questions can no longer be illuminated by conceivable experiments'. His conclusion? 'It's a very disquieting position to be in'.[45]

Never one to shirk a challenge, Weinberg struck out on two courses simultaneously: to push for the building of a huge collider machine underground in Texas, and to narrow down the search for the ultimate theory of nature using the *Quest for Beauty* as guide. The book that issued from this twin-track approach, *Dreams of a Final Theory* (1993), sought support for the Superconducting Super Collider by outlining an aesthetics-based methodology as a guide, or programme, for scientific development.

The collider is now dead, as we have seen. But Weinberg's dreams of a theory to end all theories are equally a symptom of their times. His assumption is that there are a finite number of fundamental laws of nature, and that these must in some sense fit together. In other words, Weinberg believes natural law has a *simplicity*, and an *inevitability*: a simplicity in that the number of laws is finite, and an inevitability in that the nature of one law must relate to the whole, and be constrained by the nature of other laws.

These two criteria, Weinberg takes to define beauty: 'the beauty of perfect structure, the beauty of everything fitting together, of nothing being changeable, of logical rigidity'.[46] Yet both simplicity and inevitability can, in turn, be captured by principles of *symmetry*. That is why the principle of symmetry is popular with many particle physicists and others.

Weinberg holds that, if we find that our theories of nature possess this beauty, we must really be on the right road to the End of Physics. In fact scientists, he argues, are *already* 'learning how to anticipate the beauty of nature at its most fundamental level'. Therefore Weinberg is in a mood to celebrate the possibility of an early implementation of his aesthetic programme for science. Of current aesthetically-orientated opinion in science, he concludes: 'nothing could be more encouraging that we are actually moving toward the discovery of nature's final laws'.[47]

At least Weinberg believes in the possibility of expanding human knowledge. That alone marks him off from the mainstream catatonia that bedevils contemporary science. There are, nevertheless, some important weaknesses in his argument.

It is rather circular. Nature, we are told, is simple and interconnected. Beauty is simplicity and inevitability. Therefore, beauty is a guide to nature's laws.

Weinberg is more of a rational materialist than many of his peers, but he cannot argue why, still less prove that, nature is ordered in an attractive way. He *believes* that it is thus structured. Eventually, Weinberg chases beauty with the best of them:

> Plato and the neo-Platonists taught that the beauty we see in nature is a reflection of the beauty of the ultimate, the *nous*. For us, too, the beauty of present theories is an anticipation, a premonition, of the beauty of the final theory. And in any case, we would not accept any theory as final unless it were beautiful.[48]

Here, science really does seem to gain the mantle of a substitute religion.

In an earlier book, *The First Three Minutes*, Weinberg contended that the universe is pointless. He wrote, too, that the more we know about the universe, the more pointless we find it to be. At the end of *Dreams of a Final Theory*, by contrast, he says he feels 'sad' that the universe has no meaning. In the *Quest for Beauty*, it seems to us, he has found a brightly-lit kind of consolation.

The Quest for Beauty as a Problem for Science

The *Quest for Beauty* is not just a little harmless opium for theoretical physicists. Nor, sadly, is it just an interesting cultural expression of wider social trends as they are reflected in the physics community. Like the Copenhagen interpretation of quantum mechanics, and the belief that chaos and complexity theories provide a universal theory of nature, the aesthetic conception of nature poses dangers for science.

Weinberg's approach might make sense if there was signif-icant evidence that his mathematical laws captured something fundamental about the nature of physical reality. But there is precious little evidence of this. In fact, Weinberg admits to fiddling things so as to get the beauty he desires: 'Our principles', he admits, 'are often invented as we go along, sometimes precisely because they lead to the kind of rigidity we hope for'.[49]

To reach the kind of rigidity he seeks, Weinberg, like others, paradoxically offers theories so flexible, so far removed from experimental verification, that a wide range of them could fit the data. Which theory triumphs is, therefore, largely a matter of taste. In the contemporary context, which is that of an outpouring of theories unmatched by experimental advances, *subjective idealism* can all too easily subvert science as a rational process.

In the past, experiments played a vital role in developing theory. Today, experiments in some fields are barely managing to test out theories developed a decade or more ago. Wherever experimental evidence can be coaxed out of nature, it suffices to corroborate or refute a theory and serves as the sole arbiter of validity. But where evidence is spare or absent – as it is for a growing number of questions in physics – other criteria, including aesthetic ones, have been allowed to come into play. Indeed, aesthetics now have moment not only in the *interpre-tation* of new theories, but also in their *formulation*.

Philosophers of science are concerned with investigating what is distinctive about the explanations and theoretical constructions of science. They are also concerned with hiving off science from guesswork – with what is called in the literature the 'demarcation problem'. Of fundamental interest to philosophers of science is the scientific method. Here, there are competing theories; the whole subject is one of increasing scholarship and intense debate. Nevertheless, most scientists, past and present, would agree that the scientific method consists of the following stages: develop a theory which marks a departure or extension from existing theory; make the theory yield experimentally testable predictions; and, lastly, put the predictions to the test of experimental evidence.

In many areas of science today, however, the equal relationship between theory and experimentation posited by the scientific method no longer holds good. Theory dominates. It is true, for example, that experimental work shared the laurels with theory in the recent discovery and development of high temperature superconductors. But, especially in particle physics and cosmology, the gap between theory and experimentation is a serious problem.

In both the subatomic domain and the realm of galaxies today, experiments and observations often take at least a decade to bring to fruition and publication. In some cases, lead-times look like proving much longer; and certainly delays allow theories which are often first advanced as purely speculative ideas to become part of the accepted wisdom for physicists. Over a period of time, an 'elegant' set of equations can be discussed and manipulated enough for them to become 'compelling'. As the cost of both colliders and ever more detailed sightings of the universe spirals, so beauty has filled the gap between theory and experimentation.

Our aesthetes have become so attached to their ideas, in fact, that they are unwilling to jettison them even when experimental data is available which confounds their beautiful theories. When observational data goes against analysis, they prefer merely to modify the latter. The British science writer Robert Oldershaw has reminded us of the tradition in which today's generation of aesthetically-inclined scientists stand:

An irreverent name for this strategy might be the Ptolemaic method. Remember Ptolemy, the Greek astronomer-mathematician who, having accepted as fact the hypothesis that the Earth was the centre of the Universe with the Sun, Moon and planets revolving round the Earth, developed a theory of planetary motion that involved adding increasingly complicated 'epicycles' until his predictions fitted the facts.[50]

The satirical thrust here is obvious enough, but Oldershaw's arrows do hit their target.

In certain instances, theories are so far removed from the observable world that they can accommodate an astonishingly wide variety of experimental results. The theories, after all, have many arbitrarily adjustable parameters; framed broadly enough, they can handle any amount of data. Again, although such theories do make predictions, they are compromised by their excessive flexibility. By twiddling parameters, or switching to a slightly modified version of the theory, any possible disagreement with observational data can be overcome.

Moving the Goalposts in Particle Physics

The Standard Model of particle physics involves no fewer than 17 arbitrarily adjustable parameters. It holds that matter consists of numerous fundamental particles; also, that there are four fundamental forces between particles – strong, electromagnetic, weak and gravitational. These four forces, whose actions are themselves mediated by particles, are viewed as different manifestations of one underlying force. An analogy makes the idea clearer: ice, water and steam are just different manifestations of the H_2O molecule.

There are precedents in physics for the unification of what appear to be disparate phenomena. In the nineteenth century, James Clerk Maxwell successfully showed that electricity, magnetism and light were just different aspects of the same underlying phenomenon – electromagnetism. In the years 1967–68, Weinberg at Harvard and Abdus Salam at Imperial

College, London, independently proposed a theory which showed that the electromagnetic and weak forces were just different aspects of one force – the electroweak.[51] However, despite its successes, and its antecedents in previous attempts at unification, serious problems beset the Standard Model. It does not, by itself, explain why particles have masses. It can be adjusted so that particles are assigned masses. But for this, the Standard Model must allude to the existence of what is termed a Higgs boson. And at present the Higgs boson remains undetected in even the world's largest particle accelerators.

There are those who object to the ad hoc way in which the Higgs mechanism was introduced. Yet the Model does not work without it. Worse, the Model also fails to predict the relative strengths of the forces as they manifest themselves. In sum, as Martinus Veltman has written:

> The only legitimate reason for introducing the Higgs boson is to make the Standard Model mathematically consistent... The biggest drawback of the Higgs boson is that so far no evidence of its existence has been found. Instead, a fair amount of indirect evidence already suggests that the elusive particle does not exist. Indeed, modern theoretical physics is constantly filling the vacuum with so many contraptions that it is amazing a person can even see the stars on a clear night![52]

It is no wonder that Princeton physicist James Peebles was moved to observe, of contemporary research, that much of it was 'fed by the thin gruel of theory and negative observational results, with no prediction and experimental verification of the sort that, according to the usual rules of evidence in physics, would lead us to think we are on the right track'.[53]

The tendency for physicists to get entangled with metaphysics, by moving the goalposts for what a theory is meant to capture, is also very clear with superstring theorists. About this group, Sheldon Glashow, a leading theoretical physicist and Nobel Prize winner, has argued:

Some of them are convinced in the uniqueness and beauty, and therefore truth, of their theory, and since it is unique and true it obviously includes a description of the entire physical world. It does not seem to them to be necessary to do any experiments to prove such a self-evident truth, so they begin to attack the value of experiments from this end – a highly theoretical, abstract, mathematical end.[54]

Richard Feynman expressed similar concerns:

I don't like that they're not calculating anything. I don't like that they don't check their ideas. I don't like that for anything that disagrees with an experiment, they cook up an explanation – a fix-up to say 'well, it still might be true'.[55]

Superstring theory is highly abstract and difficult to comprehend. It has provided few specific predictions. The predictions it *has* made have been contradicted by the available experimental evidence. The masses of particles predicted by the theory do not correspond to the known masses. However, superstring theorists argue that, as the theory is developed, it will produce the corrections that are required to give the right masses. Nor are they deterred by the fact that no particles have ever been seen that have the energies predicted by the theory.

Existing particle accelerators cannot produce particles at the high energies suggested by superstrings. So superstring theory has not made contact with the real world of the experimental physicist. However, it presently detains most of the world's best theoretical physicists. Weinberg has remarked on this contrast:

The intellectual investment now being made in string theory without the slightest encouragement from experiment is unprecedented in the history of science. Yet for now, it offers our best hope for a deeper understanding of the laws of nature.[56]

'Best hope' or not, the intellectual energy being devoted to such a speculative theory sets a dangerous precedent.

Some years ago, it was put to Sheldon Glashow that theories which attempt to unify nature in the abstract superstring manner might threaten the future of physics. With the idea that such theories undermine the motivation to experiment, Glashow concurred:

> Yes, in the same way that I think medieval theology destroyed science in Europe in the middle ages. It was, after all, only in Europe that people did not see the great supernova of 1054, for they were too busy arguing how many angels could dance on the head of a pin!... I think that the old tradition of learning about the world by looking at the world will survive, and we will not succeed in solving the problems of elementary particle physics by the power of pure thought itself.[57]

Glashow is surely right. But the fact is that the *Quest for Beauty* is currently the mainstream position in particle physics.

Science and Art

So far, we have had pause to comment on the affinity between aesthetics in science on the one hand, and religion on the other. Yet it is also the case that, once people accept 'beauty' as a guide, they tend to obliterate the distinction between natural science and the arts. Indeed, this is the conclusion that some scientists and commentators this century have in fact drawn – that science is akin to art.

In an interview, the theoretical physicist Michael Berry was asked: 'You keep talking about the aesthetics. Can you explain what you mean when you speak about the "elegance" of the theory?'. 'Very difficult' was Berry's first response. But he went on:

> Taste is a difficult concept to define and it's something that's appreciated by people who know it and not by people

who don't. I don't want to sound terribly aristocratic, I certainly don't feel that way. You can only explain it by analogy. A piece of music, let's say, can be ill-constructed or well-constructed, and you can after a while hear – learn to hear – the difference. It's like that with theories. I wouldn't like to give the impression that there could be no objective way of assessing the aesthetic value of a physical theory, but I don't know one.[58]

Just to make sure that he was not misunderstood, Berry added that 'you could not communicate it' [the 'difference'] 'to somebody who doesn't share the feeling. I don't think that's possible'.[59]

Berry's view is similar to that expressed by Dirac. For Dirac, mathematical beauty could not be defined 'any more than beauty in art can be defined'. At the same time, he maintained, people who study mathematics usually have no difficulty in appreciating that beauty when they see it.[60] J. W. N. Sullivan, the author of biographies of both Newton and Beethoven, took a similar view back in May 1919:

Since the primary object of a scientific theory is to express the harmonies which are found to exist in nature, we see at once that these theories must have an aesthetic value. The measure of the success of a scientific theory is, in fact, a measure of its aesthetic value, since it is a measure of the extent to which it has introduced harmony in what was before chaos.

It is in its aesthetic value that the justification of the scientific theory is to be found, and with it the justification of the scientific method. Since facts without laws would be of no interest, and laws without theories would have, at most, only a practical utility, we see that the motives which guide the scientific man are, from the beginning, manifestations of the aesthetic impulse.[61]

For much of the twentieth century, the heroes of science have been portrayed as creative geniuses, real heroes in their field. But such a romanticisation of the individual scientist was always suspect.

The art critic Roger Fry was ready, despite his romantic inclinations, to challenge Sullivan:

> I should like to pose... the question whether a theory that disregarded facts would have equal value for science with one which agreed with facts. I suppose he [Sullivan] would say No; and yet so far as I can see there would be no purely aesthetic reason why it should not.[62]

Fry was right: an appeal to aesthetics alone is not enough to distinguish between one theory and another. The real test comes when the predictions of each are checked against experimental observations.

What, though, happens when, as Hawking's says, a theory is 'put forward for aesthetic or metaphysical reasons', and we have no significant way of empirically testing it? The temptation arises to accept as science what is little more than aesthetic judgement. And that temptation is made all the grislier in its results where it is recalled that aesthetic judgements are those of individuals. As the Scottish philosopher David Hume said: 'Beauty in things exists in the mind which contemplates them'.[63] Also, aesthetic sensibilities change over time. For example: what would a Kepler have made of Fiegenbaum's view – that fractal shapes are beautiful? Considerations such as these should warn against any claim that beauty can be given a level of objectivity appropriate to the study of nature.

The gap between theory and experiment means that there are, today, no scientific alternatives to superstrings and similar inventions. The gap is responsible both for the problems of theoretical physics, and for the fact that speculation has now become the best shot in many physicists' lockers. About this there can be no argument. Let it be clear, too, that speculation is a crucial part of scientific progress. It must be encouraged. Nor do we deny that beauty itself is also a wonderful thing to behold. Just as it is only human for scientists to gain satisfaction from flights of fancy, so they like to construe beauty in nature. Historically, as we have seen, Kepler moved from beautiful schemas to mundane ellipses; but, using beauty as a guide, individuals have, from time to time, made durable contributions to science.

The fun, recreation and pleasure that is to be had from inquiring into boundless nature – all this is not at issue. The issue is, rather, that scientists should not make a virtue out of necessity by elevating 'beauty' into a guiding principle of research. Or, if they want to do that, they should freely admit that they are engaged in mathematical pastimes, not theoretical physics. The failure to do this – the tendency to engage in the *Quest for Beauty* as if it were theoretical science – puts into jeopardy the whole rational basis of science.

Mathematics and the Loss of Certainty

Those engaged in the *Quest for Beauty*, and those who hold to the universal claims of chaos and complexity theories, both tend to substitute mathematical speculation for natural science. From this they infer that truths about mathematics are truths about nature. A striking result of this is that a theorem developed in 1931 by the logician Kurt Gödel is widely taken to undermine the possibility, not just of a Theory of Everything, but of any kind of fundamental scientific knowledge.

John Barrow has made this point very widely. In *Theories of Everything: The Quest for Ultimate Explanation* (1991), he argues that, if the world is fundamentally mathematical,

> any limitations of mathematical reasoning, like those uncovered by Gödel, are thus not merely limitations on our mental categories but intrinsic properties of reality and hence limitations upon any attempt to understand the ultimate nature of the universe.[64]

John Casti, of the Santa Fe Institute, leans in the same direction.

For him, Gödel's result puts the final nail in the coffin of scientific certainty. After laying out the idea that quantum theory and chaos theory exclude the possibility of fundamental knowledge, Casti contends that Gödel's result 'eliminates once and for all the hope of ever attaining perfect scientific prediction and explanation of anything'.[65]

In our view, Barrow and Casti are wrong on two counts. First, fixed mathematical formulae do not amount to natural science. Second, even if it were indeed the case that the world is intrinsically mathematical, it would not follow that ignorance is inescapable. What Barrow and Casti betray, on close inspection, is another species of belief: belief that Gödel's result *must* place limits on human understanding. Because it is argued from first principles of logic, it is this doctrine which forms, perhaps, the clearest expression of today's *Loss of Certainty* in science. With it, therefore, we conclude this book.

It is striking that Barrow and Casti make stronger claims than Gödel himself made for his results. In a passage which Barrow can only allude to as 'strange', Gödel argued that, when faced with uncertainty, humanity could always seek to expand knowledge by developing new axiomatic systems. Referring to a conjecture made by the mathematician Georg Cantor [1845–1918], Gödel wrote that 'its undecidability from the axioms as known today can only mean that these axioms do not contain a complete description of reality'.[66] For Gödel, incompleteness in any particular theory was a challenge, not a way of life.

What, though, is Gödel's theorem exactly, and what consequences does it have? Morris Kline gives this summary:

If any formal theory T adequate to embrace number theory is consistent and if the axioms of the formal system of arithmetic are axioms or theorems of T, then T is incomplete. That is, there is a statement S of number theory such that neither S nor not-S is a theorem of the theory. Now either S or not-S is true; there is, then, a true statement of number theory which is not provable... One of the implications of Gödel's theorem is that no system of axioms is adequate to encompass, not only all of mathematics, but even any one significant branch of mathematics, because any such axiom system is incomplete... Gödel's theorem dealt a death blow to comprehensive axiomisation.[67]

Mathematicians can, of course, seek to extend their axioms in order to make S or not-S a theorem of the theory. But then if their new system of axioms is consistent, there will be some other statement – call it P – such that neither P nor not-P is a theorem of the theory.

After 1931, most mathematicians decided that Gödel's result put an end to the great attempts to put maths on a rigorous axiomatic footing – attempts led by Russell, Whitehead, and David Hilbert. Today this seems indisputable; but what does it imply for physics?

Stripped of the technicalities, Gödel's result is not difficult to comprehend: any formal system has its limitations. Yet the result says more than this. We will always have to suffer limitations directly, because there will be statements which we can frame within the formal system, but which are undecidable within the system.

Gödel's result has, however, no broader consequences. It certainly does not negate the possibility of fundamental knowledge of nature. Nor does it, by itself, even rule out the possibility of a Theory of Everything.

Gödel was a Platonist. He believed that, in some sense, there exists a realm of mathematical objects which mathematicians discover. On the nature of mathematical sets, for example, he wrote:

> It seems to me that the assumption of such objects is quite as legitimate as the assumption of physical objects and there is quite as much reason to believe in their existence. They are in the same sense necessary to obtain a satisfactory theory of mathematics as physical bodies are necessary for a satisfactory theory of our sense perceptions.[68]

In holding these views, Gödel was joined by G. H. Hardy, who believed 'that mathematical reality lies outside us, that our function is to discover or observe it, and that the theorems which we prove, and which we describe grandiloquently as our "creations", are simply our notes of our observations'.[69]

Gödel and Hardy, then, distinguished between the *axioms and theorems mathematicians hold to*, and an *independent world of mathematics to be studied*. That is why Gödel, unlike Barrow and Casti, did not believe that his result set a limit to what could be known about reality. If the truth or falsity of a statement was undecidable within a given framework, there was nothing to rule out the extension of that framework to tackle the *particular* statement in question. From this perspective, Gödel's result is a *liberation* from formalism. It tells us that, rather than trying to build logically consistent systems, we should seek a closer acquaintance with reality.

Gödel and Hardy thought that mathematics was merely a discovery. Moreover, the fact that simultaneous discoveries have occurred in mathematics just as they have in physics – Newton and Leibniz's joint discovery of calculus is an obvious example – indicates that mathematics is similar to physics in having objectivity about it. But this still leaves open the precise character of that objectivity. The ability of mathematics to model material reality suggests either that everything is mathematical; or, that mathematics is a system of thought rooted in material reality. We find the latter view a much more convincing description of the relationship between maths, physics, and reality.

The former view is fundamentally Platonic. The latter, realist perspective implies something more – that mathematics is not just a discovery, nor even a delightful pastime, but very often also a human invention made to *capture* material reality.

From the realist perspective as from the Platonic perspective, mathematics is, finally, also a system of ideas and theories which is closed, which is logically coherent. In this respect there is a crucial difference between the unity that obtains in mathematics and that offered by physics. As David Ruelle puts it:

The unity of mathematics is due to the logical relation between different mathematical theories. The physical theories, by contrast, need not be logically coherent; they have unity because they describe the same physical reality.[70]

Because it is in part *invention*, and because it must also possess logical *consistency*, a mathematical model of a physical system can never be the same as the physical system itself. Einstein expressed this point strongly when he wrote: 'as far as the propositions of mathematics refer to reality, they are not certain; and as far as they are certain, they do not refer to reality'.[71]

The reverse side of the imperfect relationship between a mathematical model and physical reality is that some properties of any model follow from its nature as mathematics, not from the reality it is modelling. This is both a strength and a weakness. It is a strength because while Gödel's theorem, of necessity, applies to any logically consistent mathematical model, the questions which the model is unable to answer may not be relevant to its function as a model of reality. Here, at least, a single Theory of Everything is quite possible.

On the other hand, the intrinsically mathematical properties of a model represent a weakness. They are this because we cannot, through mathematics, have a direct understanding of physical reality. This is why some of the best physicists of the twentieth century have possessed physical intuition as well as mathematical ability.

Nevertheless, the road to physical knowledge must pass through mathematics. As Feynman wrote:

I have often made the hypothesis that ultimately physics will not require a mathematical statement, that in the end the machinery will be revealed, and the laws will turn out to be simple, like the chequer board with all its apparent complexities. But this speculation is of the same nature as the other people make – 'I like it', 'I don't like it' – and it is not good to be too prejudiced about these things.[72]

The question of whether or not a Theory of Everything is possible depends upon the character of physical reality, rather than the nature of mathematics or logic. If there is a unity to nature, if there are a finite number of natural laws, then a Theory of Everything is possible. If, by contrast, nature lacks unity, then science will have to content itself with understanding each of its disparate parts separately.

Whatever the situation in nature, of this we can be certain: the belief that Gödel's theorem places limits on human understanding speaks of a drastic *Loss of Certainty* in the thinking of many scientists.

Reversing the Retreat from Reason

For ourselves, we remain confident that humanity can comprehend the world, and can therefore reach a more revealing set of certainties about that world. Enlightenment thinkers held that this advance of knowledge pointed to a limitless 'improvement of human faculties'. Their faith in this is something which needs to be upheld against this century's crisis of faith. However, the lesson to be drawn from our bleak century is that the dream of the Enlightenment will remain nothing more than a dream without advance, not only in science, but also in society, across a broad front.

If we can tackle the barriers to progress thrown up by society, it will be possible to ground progress in more than faith, it will, in short, be possible to reverse the *Retreat from Reason*.

Notes and References

INTRODUCTION: THE SENSE OF AN END

1 Kuhn (1990) p. 21.
2 Ross (1991) p. 135.
3 *Times Higher Education Supplement*, 13 March 1992.
4 Lederman (1991).
5 Lederman (1991) p. 4.
6 Lederman (1991) p. 4.
7 In *Ecology in the Twentieth Century: A History*, her fine survey of the history of ecological movements from 1880 to the present, Anna Bramwell charts changed perceptions of nature, as well as a series of attempts to transcend nature. Until 1960, Bramwell suggests, nature was seen as the source of inequality, while attempts to transcend it through science were seen as bringing equality. From 1965 to the present, she argues, the situation has been reversed, especially among young people. 'Nature' now signifies equality; while attempts to transcend nature are seen as a source of inequality. See Bramwell (1989) p. 19.
8 Bush (1946) p. 52.
9 Cited in Pick (1989) p. 12.
10 Cited in Holton (1993) pp. 176–177.
11 George E. Brown (1992), 'Rational Science, Irrational Reality: A Congressional Perspective on Basic Research and Society', *Science*, 258: 200.
12 George E. Brown (1992), 'The Objectivity Crisis', *American Journal of Physics*, 60: 780.
13 Holton (1986) p. 283.
14 Scott (1990) p. 2.

CHAPTER ONE: THE POST-WAR LOSS OF CERTAINTY

1 Brooke (1991) p. 318.
2 Cited in Holton (1986) p. 183.
3 Descartes (1968) p. 78.

4 Westfall (1988) p. 117.
5 Cited in Shelley (1985) p. 25.
6 Cited in Hampson (1968) p. 233.
7 Russell (1990) p. 394.
8 Popper (1968) p. 32.
9 Newton–Smith (1990) p. 52.
10 Newton–Smith (1990) p. 54.
11 Popper (1986) pp. 64–5.
12 Popper (1986) p. 75.
13 Carr (1987) p. 155.
14 Kuhn (1970) p. 94.
15 Kuhn (1970) p. 206.
16 In Lakatos and Musgrave (1970) p. 20.
17 Scott (1990) p. 95.
18 Cited in Rhodes (1986) p. 734.
19 Bush (1946) p. 175.
20 See, for example, Powers (1993) and Walker (1989).
21 Kevles (1987) p. x.
22 Cited in Jungk (1958) p. 118.
23 Jungk (1958) p. 328.
24 Reeves (1991).
25 Reeves (1991) p. 8.
26 Reeves (1991) p. 31.
27 Ross (1991) p. 138.
28 Cited in McDougal (1985) p. 238.
29 Cited in McDougal (1985) p. 225.
30 Cited in Devine (1993) p. 198.
31 Cited in Hurt (1988) p. 52.
32 McDougal (1985) p. 456.
33 Scott (1990) pp. 66–67.
34 Berman (1983) p. 332.
35 Tilby (1992) p. 5.
36 Parker and Stacey (1994) p. 14.
37 Casti (1992) pp. 404–5.
38 Rich (1994) pp. 274–275.
39 Rich (1994) p. 297.

CHAPTER TWO: BELITTLING HUMANITY

1 Cited in Robertson (1905) p. 259.
2 Einstein's approach is discussed in Chapter Three. Fine (1986) provides a further source for Einstein's views. Weinberg's approach is discussed in Chapter Seven; see also Weinberg (1993).
3 For a good popular discussion of this, see Casti (1992).
4 Cited in Barrow and Tipler (1989) p. 49.
5 Cited in Gribbin (1985) p. 5.
6 Cited in Moore (1989) p. 232. In 1895 Lorentz found a way of trans-

forming electromagnetic quantities in one frame of reference into their values in another frame moving relative to the first. The full significance of this discovery was not realised until Einstein's theory of special relativity 10 years later. See Beiser (1987) p. 33.

7 Koestler (1972) pp. 138–9.
8 Cited in Gibbins (1987) pp. 4–5.
9 In Hall (1992) p. 215.
10 Cited in Pais (1991) p. 306.
11 Bohr (1958) p. 60.
12 Barrow (1990) p. 16.
13 Cited in Pais (1982) p. 399.
14 Cited in Pais (1982) p. 15.
15 Cited in Pais (1991) p. 237.
16 Hall (1992) p. 10.
17 W. J. Firth (1991), 'Chaos – Predicting The Unpredictable', *British Medical Journal*, 303: 1565.
18 Peters (1992) p. 202.
19 i–D, April 1990.
20 Some writers think that Scotland's James Clerk Maxwell preceded the French when, in 1873, he studied in some depth the problem of stability and instability. See Barrow (1990) pp. 272–4.
21 Cited in Ruelle (1991) p. 48.
22 Poincaré (1905) p. 65.
23 Poincaré (1905) p. 68.
24 Ruelle (1991) p. 48.
25 See Gleick (1988) and Bass (1990) for the history of and amusing anecdotes from this period.
26 *New Scientist*, 3 August 1991.
27 Ruelle (1991) pp. 71–2.
28 See Zurek (1990) for a survey of some of the work. A more particular approach is developed in Nicolis and Prigogine (1989).
29 For a report of his work see *Physics World*, November 1991.
30 Barrow (1991) p. 138.
31 Cited in Weber (1990) p. 182.
32 Cited in Weber (1990) p. 185.
33 Gell–Mann (1994) p. 14.
34 Reeves (1991) p. 175.
35 Davies (1993) p. 232.
36 Weber (1990) p. 194.
37 Cited in Waldrop (1992) p. 321.
38 Barrow (1992) p. 163.
39 Cited in Waldrop (1992) pp. 330–1.
40 Darwin (1985) p. 229.
41 Davies (1989).
42 Weber (1990) p. 187.
43 See Lucas (1989) and Harris (1991).

CHAPTER THREE: QUANTUM MECHANICS AND THE 'RISKY GAME'

1 Cited in Hiley and Peat (1987) p. 5.

2 Feynman (1967) p. 129. First published by the BBC in 1965.

3 Planck, M. (1901), *Annalen der Physik*, 4, 3: 553.

4 The 1918 Nobel Prize for Physics was not awarded until 1919.

5 Interview with J. Franck, *Archives for the History of Quantum Physics*, (7 September 1962): 6. American Institute of Physics, New York City. The American Physical Society and the American Philosophical Society undertook a joint project in the late 1950s and 1960s to assemble sources for the historical study of quantum mechanics. The project director was Thomas Kuhn. One of the project's main aims was to carry out interviews with the major figures responsible for the emergence and development of quantum mechanics.

6 Thomson was an influential and inspirational figure. Seven of his students went on to win the Nobel Prize.

7 Cited in Burcham (1963) p. 49.

8 Letter from Rutherford to B. Boltwood, 14 December 1910. Reprinted in Badash (1969) p. 235.

9 For the series of three papers that Bohr published in 1913 in which he presented his quantum theory of atomic structure, see Mehra and Rechenberg (1982), Chapters One and Two.

10 Bohr actually called them 'stationary states'; the term 'energy levels' was introduced later.

11 Cited in de Vaudrey Heathcote (1954) p. 415.

12 There are two types of line spectra: emission and absorption, depending upon whether photons are emitted or absorbed. In appearance, one is the negative of the other.

13 See Quine (1970).

14 Murdoch (1989) p. 17.

15 The term 'photon' first appeared in print in 1926.

16 For attempts at classical explanations of the photoelectric effect, see R. H. Stuewer, 'Non-Einsteinian Interpretations of the Photoelectric Effect', in Stuewer (1979).

17 Pais (1982) p. 382.

18 Cited in Murdoch (1989) p. 22.

19 Cited in Langevin and de Broglie (1912). The Solvay Conferences were a series of scientific meetings sponsored by the Belgian chemist Ernest Solvay. Solvay made a fortune from the manufacture of sodium carbonate. He provided funds for these meetings to continue after his death, so that leading physicists could gather and exchange ideas.

20 Millikan was awarded the 1923 Nobel Prize for physics. He had established experimentally that the electron was the fundamental unit of electric charge, and had done superb work on the photoelectric effect.

21 See R. Millikan (1949), *Review of Modern Physics*, 21: 343.

22 Cited in Arrhenius (1965) p. 478. The 1921 Prize was awarded in 1922.

23 Cited in Gillispie (1960) p. 473.
24 Cited in Speziali (1972) p. 130.
25 See Stuewer (1975).
26 S. K. Allison (1965), 'Arthur Holly Compton', *Biographical Memoirs of the National Academy of Science*, 38: 81.
27 Cited in Pais (1982) p. 414.
28 This tends to confirm the views of the American philosopher Willard Van Orman Quine. Quine argues that a scientific theory can be regarded as an interconnected web, with no part immune to revision in the light of experimentation. But nor can experimentation force the rejection of one particular part. He believes that scientists are prepared to tamper with the web at its periphery in an attempt to minimise disruption nearer the centre. The centre is occupied by the laws of physics, mathematics and logic. In many respects this echoes the earlier work of the French philosopher of science and physicist Pierre Duhem. See Quine (1970).
29 Cited in McCormmach (1970) p. 19.
30 Cited in Pais (1982) p. 414.
31 See Pais (1994), Chapter Three: 'De Broglie, Einstein, and the Birth of the Matter Wave Concept'.
32 The dual character of matter is nicely embodied in the Thomson family. J. J. Thomson was awarded the 1906 Nobel Prize for physics 'for his theoretical and experimental investigations of the passage of electricity through gases'. In effect, the citation of the Nobel Committee acclaimed Thomson for proving that electrons were particles. J. J.'s son, George, was from another school. He shared the Nobel Prize for physics in 1937 with Clinton Davisson – for work which proved that electrons behaved like waves. They used crystal lattices to diffract electrons.
33 Jammer (1966) p. 196.
34 An English translation of the paper can be found in van der Waerden (1967).
35 Van der Waerden (1967) p. 261.
36 Interview with Thomas Kuhn in 1963; cited in Folse (1985) p. 84.
37 Einstein was in Berlin from 1914 to 1933. After Hitler came to power, he spent the rest of his life at The Institute for Advanced Study at Princeton.
38 Heisenberg was awarded the 1932 Nobel Prize (actually decided in 1933) for 'his establishment of quantum mechanics whose application has led, among other things, to the discovery of the allotropic forms of hydrogen'. Schrödinger shared the 1933 prize with Paul Dirac 'for their discovery of new and productive forms of atomic theory.' See Weber (1981) pp. 95, 99.
39 See Moore (1989) Chapter Six: 'Discovery of Wave Mechanics'.
40 Cited in Cassidy (1992) p. 213.
41 Cited in Cassidy (1992) p. 215.
42 Cited in Cassidy (1992) p. 215.
43 They were shown to be equivalent by the great English physicist Paul Dirac.
44 Heisenberg (1971) p. 75.
45 The dynamical states of a quantum mechanical system (e.g. a particle)

can be described by a complex wave function. The wave function is a single-valued function of the parameters of the system, which contains all that can be known about the system.

46 For an English translation, see Wheeler and Zurek (1983) pp. 62–84.
47 The other form relates energy and time.
48 A beam of electrons could be used instead of light.
49 See J. Hilgevoord and J. Uffink, 'A New View on the Uncertainty Principle', in Miller (1990). The authors examine Heisenberg's microscope argument. They suggest that Heisenberg's argument depended on a relation between two quite distinct concepts of uncertainty. The first was an uncertainty in what was predicted; the second was related to the notion of resolving power, and was an uncertainty in what could be inferred.
50 Cited in Murdoch (1989) p. 48.
51 For the development of Bohr's ideas on complementarity, see Folse (1985).
52 Segrè (1980) p. 168.
53 N. Bohr, 'Discussions with Einstein on Epistemological Problems in Atomic Physics', in Schilpp (1949) p. 210.
54 Cited in Folse (1985) p. 159.
55 Cited in Laurikainen (1988) p. 164.
56 Davies and Brown (1986) p. 11.
57 Wheeler (1982) p. 13.
58 Wheeler (1982) pp. 11–13.
59 Wheeler (1981) p. 15.
60 Elsewhere Wheeler has written: 'May the universe in some strange sense be "brought into being" by the participation of those who participate?... The vital act is the act of participation. "Participator" is the incontrovertible new concept given by quantum mechanics. It strikes down the term "observer" of classical theory, the man who stands safely behind the thick glass wall and watches what goes on without taking part. It can't be done, quantum mechanics says.'
61 Cited in Kline (1986) p. 241.
62 Davies (1989) p. 165.
63 Letter dated 22 December 1950. Translated in Prizbam (1967) p. 39.
64 We concentrate on the Copenhagen interpretation because of the anti–Enlightenment conclusions which are drawn from it. It is also still the most influential interpretation. However, there are other interpretations. For a discussion of the 'consistent histories' interpretation, see Gell–Mann (1994), and Omnès (1994).
65 Cited in Mermin (1990) pp. 113–114.
66 Letter from Einstein to D. Lipkin, 5 July 1952. Cited in Fine (1986) p. 1.
67 Cited in Fine (1986) pp. 97–8.
68 The paper was received by the editors of the *Physical Review* on 25 March 1935.
69 Alain Aspect et al (1982), 'Experimental Realization of Einstein–Podolsky–Rosen–Bohm *Gedankenexperiment*: A New Violation of Bell's Inequalities', *Physical Review Letters*, 49: 91.
70 Niels Bohr (1936), 'Can the Quantum Mechanical Description of Physical Reality be Considered Complete?', *Physical Review*, 48: 697.

71 See A. Shimony, 'An Exposition of Bell's Theorem', in Miller (1990).
72 Davies and Brown (1986) p. 52.
73 In Davies and Brown (1986) p. 51.
74 In Davies and Brown (1986) p. 54.
75 Cushing (1994) p. 39.
76 A. J. Leggett, 'Reflections on the Quantum Measurement Paradox', in Hiley and Peat (1991) p. 85.
77 Einstein (1950) p. 88.
78 Cited in Pais (1982) p. 515.
79 Cited in Squires (1990) p. 180.
80 Cited in Pais (1991) pp. 426–7.
81 Cited in Fine (1986) p. 95.
82 Heisenberg (1989) p. 116.
83 Wolfgang Pauli to Max Born, in Born (1971) p. 223.
84 Heisenberg (1989) p. 32.
85 Cited in Cassidy (1992) p. 250.
86 Letter from Einstein to Schrödinger 31 May 1928. Reproduced in Pirzbam (1967) p. 31.

CHAPTER FOUR: CHAOS, COMPLEXITY AND CONTROL

1 Wolfram is a polymath with wide interests in mathematics, physics, and computing. Among other things, he helped develop the theory of cellular automata, and wrote the computer program *Mathematica*.
2 Lorenz (1993) p. 8.
3 Reprinted as an appendix in Lorenz (1993).
4 Casti (1992) p. 74.
5 See Stewart (1989) pp. 104–105.
6 Casti (1992) p. 75.
7 Davies and Gribbin (1991) p. 35.
8 Gleick (1988) p. 304.
9 Baker and Gollub (1990) p. vii.
10 Casti (1992) pp. 75–76.
11 Gleick (1988) p. 17.
12 Ruelle (1991) pp. 62–63.
13 Coveney and Highfield (1990) p. 245.
14 Coveney and Highfield (1990) pp. 221–2.
15 Coveney and Highfield (1990) p. 209.
16 Parker and Stacey (1994).
17 Peters (1992) p. 109.
18 Cited in Hall (1992) p. 180.
19 In Zurek (1990) p. 63.
20 In Zurek (1990) pp. 63–64.
21 See for example Hall (1992), Roderick V. Jensen, 'Quantum Chaos', *Nature*, 355: 311, Martin C. Gutzwiller, 'Quantum Chaos', *Scientific*

American, January 1992; Gell-Mann (1994) pp. 26–27, and Cohen and Stewart (1994) p. 236. Gell-Mann argues that chaos interacts with quantum to *amplify* quantum indeterminacy. Jensen and Gutzwiller argue that chaos does in fact penetrate the quantum domain. But Gutzwiller says that he 'must emphasise, however, that the term "quantum chaos" serves more to describe a conundrum than to define a well-posed problem.' Jensen argues for a radical revision of quantum mechanics so as to make it compatible with chaos. Cohen and Stewart seem more to the point when they argue that the discreteness of quantum mechanics should abolish chaos.

22 For some examples of how chaos can be used, see William L. Ditto and Louis M. Pecora, 'Mastering Chaos', *Scientific American*, August 1993.
23 In Zurek (1990) p. 61.
24 Prigogine and Stengers (1984) p. 22.
25 Nicolis and Prigogine (1989) p. 73.
26 Nicolis and Prigogine (1989) p. 56.
27 Cited in Briggs and Peat (1989) p. 146.
28 Prigogine and Stengers (1984) p. 9.
29 See McGuinnes (1974).
30 Weinberg (1993) p. 34.
31 Gell-Mann (1994) p. 371.
32 Cited in Waldrop (1992) p. 235.
33 Cited in Lewin (1993) p. 184.
34 Kauffman (1993) p. 644.
35 Kauffman (1993) p. 642.
36 Kauffman (1993) p. 645.
37 Kauffman (1993) p. 407.
38 Kauffman (1993) p. 407.
39 Kauffman (1993) p. 409.
40 Cited in Waldrop (1992) p. 321.
41 Kauffman (1993) p. 239.
42 Kauffman (1993) p. 261.
43 Kauffman (1993) p. 173.
44 Kauffman (1993) p. 522.
45 Waldrop (1992) p. 313.
46 Kauffman (1993) p. 280.
47 Kauffman (1993) p. 595. John Maynard Smith believes that such evidence as exists on development in Drosophila counts more against Kauffman than for him (Maynard Smith, personal communication).
48 Kauffman (1993) p. 644.
49 Kauffman (1993) p. 464.
50 Kauffman (1993) p. 239.
51 Kauffman does discuss the possibility of letting N vary, but does not follow through to examine the consequences. He claims that an increasing N will, beyond a certain point, become more of a hinderance than a benefit. This is simply stated rather than shown, however. See Kauffman (1993) p. 54.
52 Gabriel A. Dover (1993), 'On The Edge', *Nature*, 365: 704.

53 Gould (1990) p. 51.
54 Gell-Mann (1994) p. 323.
55 Gould (1990) p. 11.
56 Gould (1990) p. 10.
57 Prigogine and Stengers (1984) p. 4.
58 Rose (1992) p. 150.
59 Rose (1992) p. 150.
60 Lewontin (1993) p. 121.
61 Cited in Gleick (1988) p. 153.
62 Coveney and Highfield (1990) p. 206.
63 Cited in Hall (1992) p. 59.
64 Ruelle (1991) p. 74.
65 Stewart (1989) p. 191.
66 Parker and Stacey (1994) p. 63.
67 Gell-Mann (1994) p. 321.
68 Whitehead (1985) p. 90.
69 Whitehead (1985) p. 218.
70 Whitehead (1985) p. 222.
71 Chardin (1959) p. 107.
72 Chardin (1959) p. 297.
73 Jantsch (1992) p. 308.
74 Hayek (1990) p. 9.

CHAPTER FIVE: SCIENCE AND HUMANISM

1 Butterfield (1951) p. 174.
2 Einstein's work successfully overcame limitations in Newton's. Nevertheless, Einstein paid his own tribute:

Newton, forgive me; you found the only way which, in your age, was just about possible for a man of the highest thought and creative power. The concepts, which you created, are even today still guiding our thinking in physics, although we now know that they will have to be replaced by others farther removed from the sphere of immediate experience, if we aim at a profounder understanding.

Cited in Ferris (1988) p. 103.
3 It was written in Latin, and was not available in English until 1729.
4 Cited in Ferris (1988) p. 103.
5 Cited in Hankins (1985) p. 1.
6 For a useful discussion of the social organisation of the Enlightenment, see Russell (1983) Chapter Five.
7 See Manchester (1992).
8 See Barrow (1992) for a brief history of this development.
9 The French combined the verb *renaitre*, 'revive', with the feminine noun *naissance*, 'birth', to form renaissance – meaning rebirth. In Italy, the movement was known as the *Rinascimento*. For an introduction to the Medieval and Renaissance worlds, see Manchester (1992).
10 See Jacobs (1988).

11 J. Fernal published his work *De Abditis Rerum Causis* in 1548. Quoted in Boas (1970) p. 13.

12 Kline (1987) pp. 46–49.

13 See Dijksterhuis (1961) for more examples like this.

14 For a very useful analysis of this, see Chapter One of Rubin (1979). Rubin argues that the 'regional economy was based on a combination of the rural feudal *demesne* with the guild handicrafts in the towns; it was, therefore, only with the decomposition of both of these that the disintegration of the regional economy could occur. In both cases their decomposition was brought about by one and the same basic set of causes: the rapid development of a *money economy*, the expansion of the *market*, and the growing strength of *merchant capital*. (p.20.)

15 Mason (1962) p. 138.

16 See Hooykaas (1984) p. 47.

17 Gillispie (1960) p. 59. Emphasis added.

18 See Komesaroff (1986).

19 Kline (1987) p. 246.

20 Cited in Marks (1983) p. 238.

21 Wolpert (1992) p. 35.

22 Cited in Lindberg and Westman (1990) p. 228.

23 Galilei (1953) p. 328.

24 Cited in Fauvel (1989) p. 185.

25 See Manchester (1992) pp. 106–117.

26 For an account of the Scientific Revolution in Britain, and of Voltaire's sympathetic stance to it, see Spadafora (1990).

27 Cited in Marks (1983) p. 92.

28 Cited in Behrens (1967) p. 123.

29 Westfall (1988) p. 30.

30 Cited in Lindberg and Westman (1990) p. 245.

31 Cited in Hampson (1968) p. 36.

32 Hampson (1968) p. 149.

33 Appleyard (1992) p. 78.

34 Cited in Kline (1987) p. 294.

35 Cited in Solomon (1990) p. 8.

36 Carr (1990) pp. 160–161.

37 See Füredi (1992).

38 See Duhem (1977) for an exposition of his views.

39 Lindberg and Westman (1990) p. xviii.

40 Lindberg and Westman (1990) p. xx.

41 For the Enlightenment, Aristotelian rationalism could and had to be supplanted in one of two ways: by adopting an alternative form of rationalism, or by taking up cudgels on behalf of empiricism. By and large, the rationalists proposed that human beings are in possession of innate ideas and that, being aware of the logical relationships between these ideas, they consequently have apriori knowledge, knowledge which the rationalists claim concerns the world *as it really is*. The empiricists, on the other hand, rejected both the older Greek-Medieval doctrine of the rational intuition of the forms and the modern doctrine of innate ideas. An exception is

Kant, who did not reject the possibility of innate ideas.
42 As an example, see Brooke (1991).
43 See Brooke (1991).
44 Lukács (1980) p. 128.
45 See Chapter Seven of Füredi (1992) for more details of this.
46 There were, of course, other reactions in this period; notably the Romantic movement, which included poets and some scientists. While the scientific theories developed by people associated with this rather diffuse movement have generally proved false, and are best forgotten, it is too simplistic to label its attitudes as reactionary or conservative. There was certainly a backward-looking strain to much of Romantic poetry and writing at the time, but there was also a certain humanism. Crudely stated, the Romantics articulated a disappointment with the failure of modern society to live up to the promise of Enlightenment. True conservatives, by contrast, feared the instability which might issue from making Enlightenment promises on which society could not deliver.
47 See Jennings (1985) p. 12.
48 Cited in Lock (1985) p. 7.
49 Cited in Zeitlin (1981) pp. 59–60.
50 Hobsbawm (1988b) p. 314.
51 Pick (1989) p. 138.
52 Cited in Zeitlin (1981) p. 84.
53 See Russel (1990) pp. 697–8, for example.
54 See Landes (1989) for some excellent information on the rapidity of change in the early nineteenth century. For example, in the 1780s, Britain's output of iron was less than that of France. By 1848, Britain produced more iron than the rest of the world combined (p. 95).
55 In this sense, David Knight's conception of nineteenth century science is quite right: Science, he writes, was 'not merely an intellectual activity, but also a practical and a social one: an agent, as prophesied, in changing society.' (See Knight (1988) p. 9).

CHAPTER SIX: SCIENCE AND THE RETREAT FROM REASON

1 Schrödinger (1961) p. 3.
2 Kline (1987) p. 325.
3 Toulmin (1990) p. 168.
4 Jonathan Piel, 'Challenges for 1994', *Scientific American*, December 1993.
5 In his influential and deeply pessimistic essay entitled 'The Coming Anarchy', Robert Kaplan envisages the world slipping into a state of civil and international conflict as a growth in the number of people on the planet exceeds both the available resources of the world, and the capacities of civilisation. Kaplan and his message also formed the conclusion of a BBC *Panorama* Programme – 'Pulp Future' – screened as a part of the second national British Science Week.

6 Handy (1994) pp. 17–20.
7 Carr (1987) p. 38.
8 Carr (1987) p. 43.
9 For interesting material on this see McClelland (1970); Nye (1975); Soffer (1978); Pick (1989) and Hughes (1979).
10 Nye (1975) p. 20.
11 Nye (1975) p. 95.
12 Nye (1975) p. 112.
13 Hughes (1979) p. 41.
14 Quoted in Holton, G. (1993), 'Can science be at the centre of modern culture?', *Public Understanding of Science*, 2: 302.
15 The attitude of the Nazis to modern science is covered excellently by Cassidy (1992).
16 Cited in Nye (1975) p. 23.
17 In Gerth and Mills (1977) p. 146.
18 Freud (1991) p. 276.
19 Cited in Monk (1991) p. 485.
20 Monk (1991) p. 354.
21 Cited in Ross (1991) p. 141.
22 This point is discussed in Füredi (1992). For a discussion of eugenics in America and Germany, see Gould (1981) and Kevles (1985).
23 Scott (1990) p. 68.
24 Smith (1991) and Poudstone (1993) are fascinating and complementary histories of RAND. The former sets RAND in the context of American history and the history of American thinktanks in particular. The latter concentrates more on the science and military thinking of RAND itself.
25 In the words of Waren Weaver at a RAND conference in 1947. Cited in Smith (1991) p. 117.
26 Quoted in Poudstone (1993) p. 143.
27 Bernal (1969) pp. 1305–6.
28 Cited in Biquard (1965) p. 108.
29 Cited in Biquard (1965) p. 124.
30 Wilkie (1991) p. 102.
31 Adorno and Horkheimer (1989).
32 Adorno and Horkheimer (1989) p. 3.
33 Adorno and Horkheimer (1989) p. 9. Our emphasis.
34 Adorno and Horkheimer (1989) p. 193.
35 Carson (1982) p. 239.
36 Carson (1982) p. 257.
37 Marcuse (1972) p. 138.
38 Marcuse (1972) p. 197.
39 Marcuse (1972) p. 74.
40 Marcuse (1972) p. 130.
41 Koestler (1986) p. 53.
42 Roszak (1971) p. 51.
43 Gillispie (1990) pp. xvi–xvii.
44 Bramwell (1994) p. 60.
45 Altvater (1993) p. 216.

46 Altvater (1993) p. 80.
47 Hardin (1993) p. 56.
48 Altvater (1993) p. 198.
49 Hardin (1993) p. 45.
50 Altvater (1993) p. 202.
51 Cited in Kern (1983) p. 105.
52 Altvater (1993) p. 187.
53 Kennedy (1994) p. 330.
54 Kennedy (1994) p. 331.
55 Kennedy (1994) p. 346.
56 See Harrison (1993) p. 46.
57 Kennedy (1994) p. 15.
58 Lederman (1991) p. 6.
59 Lederman (1991) p. 11.
60 'Another anniversary to celebrate', *Nature*, 372 (1994): 1.
61 Timothy Ferris, 'The Case Against Science', *New York Review of Books*, 13 May 1993.
62 See Wolpert (1992), and John Maddox (1994), 'Defending science against anti–science', *Nature*, 368: 185.
63 Lederman (1991) p. 17.
64 'Nothing is Unthinkable', *The Lancet*, 15 September 1990.
65 Appleyard (1992) p. 256.
66 Midgley (1992) p. 224.
67 F. R. Leavis, 'The Significance of C. P. Snow', *The Spectator*, 9 March 1962.
68 'The Edge of Ignorance', in *The Economist*, 16 February 1991.
69 Duhem (1977) p. 282.
70 Duhem (1977) p. 16.
71 Duhem (1977) pp. 9–10.
72 Duhem (1977) p. 18.
73 Duhem (1977) p. 38.
74 Polanyi (1969) p. 266.
75 Polanyi (1969) p. 384.
76 Polanyi (1969) p. 405.
77 Polanyi (1969) p. 266.
78 John Wettersten (1993), 'The Sociology of Scientific Establishments Today', *British Journal of Sociology*, 44, 1: 80.
79 Barry Barnes and David Bloor are the most obvious exponents of this approach. See also Collins and Pinch (1993).
80 Hakfoort, C (1991), 'The Missing Synthesis in the Historiography of Science', *History of Science*, 29(2), 84: 207.
81 Polanyi (1969) p. 245.
82 Norman Macrae, 'Not Too Many Babies, Just Too Many Oldies', *The World in 1995*.
83 Cited in the *Independent*, 3 January 1994.
84 L. A. Friedler, quoted approvingly in Stephen G. Post (1991), 'Selective Abortion and Gene Therapy: Reflections on Human Limits', *Human Gene Therapy*, 2: 232.

85 Holton (1993) pp. 151–152.
86 Holton (1993) p. 149.
87 Cited in Gill (1986) pp. 2–3.

CHAPTER SEVEN: THE LOSS OF CERTAINTY AND THE QUEST FOR BEAUTY IN SCIENCE

1 Jammer (1966) pp. 166–7.
2 Jammer (1966) p. 198.
3 In Chant and Fauvel (1980) p. 269.
4 Cited in Chant and Fauvel (1980) p. 274.
5 Cited in Chant and Fauvel (1980) p. 274.
6 In Chant and Fauvel (1980) p. 268.
7 In Chant and Fauvel (1980) p. 294.
8 Graham (1981) p. 47.
9 In Chant and Fauvel (1980) p. 282.
10 Cushing (1994) pp. 121–22.
11 John Horgan, senior writer on *Scientific American*, picks up on this as far as complexity is concerned in his article for that magazine: 'From Complexity to Perplexity', June 1995. As the sub-head runs: 'Can science achieve a unified theory of complex systems? Even at the Santa Fe Institute, some researchers have their doubts'.
12 Cited in Waldrop (1992) p. 313.
13 Levy (1992) p. 9.
14 Kellert (1994) p. 154.
15 Kellert (1994) p. 156.
16 'The Quest for a Theory of Everything Hits Some Snags', Science (1992), 256: 1518–1519.
17 *Scientific American*, December 1992.
18 *Scientific American*, December 1992.
19 Barrow (1991) p. 150.
20 *Sunday Times*, 28 June 1992.
21 Hawking (1988) p. 136.
22 Hawking (1988) p. 136.
23 Penrose (1989) p. 152.
24 Penrose (1989) p. 155.
25 Cited in 'The Quest for a Theory of Everything Hits Some Snags', *Science* (1992), 256: 1518–1519.
26 Cited in 'The Quest for a Theory of Everything Hits Some Snags', *Science* (1992), 256: 1518–1519.
27 Interview with Witten, in Davies and Brown (1988) p. 102.
28 Penrose (1989) p. 438.
29 Cited in Heisenberg (1971). Also see Dirac (1963).
30 'The Excellence of Einstein's Theory of Gravitation', in Goldsmith, Mackay and Woudhuysen (1980) p. 44.
31 Cited in Chandrasekhar (1990) p. 65.

32 Cited in Chandrasekhar (1990) p. 52.

33 Cited in Westfall (1988) pp. 6–8.

34 Cited in Westfall (1988) p. 158. A number of authors have argued a rather different position, namely that the search for beauty and harmony played an *essential* role in the development of modern science. Easlea, for example, writes: 'No matter how difficult it may be to give operational definitions of such concepts as simplicity, harmony and beauty (or, for that matter, such concepts as love and justice), nevertheless it appears undeniable that the search for simplicity, harmony and beauty was an extremely important, perhaps essential, factor in the rise of modern science. Indeed, if men had not committed themselves to this quest, perhaps we would still be puzzling over the epicycles of the Ptolemaic planetary system.' See Easlea (1980) p. 56.

Easlea is referring to the well-known fact that Copernicus put forward his heliocentric model of the solar system on aesthetic grounds rather than scientific ones; at the time, the old Ptolemaic model fitted the observational data just as well, if not better, than Copernicus' model. But Easlea's point does not hold. After all, Copernicus' aesthetic model was replaced by Kepler's model under the influence of observational data; and, as time went by, the experimental philosophy of Newton came to dominate the thinking of the leading scientists of the day.

35 Indeed a difference between observation and prediction much smaller than that which moved Kepler to his new theory was the flaw in Newtonian mechanics which Einstein overcame with his general theory of relativity. An accurate prediction of the advance of the perihelion of Mercury was one of the major experimental confirmations of Einstein's theory. Planetary orbits tend to change their orientations in space because of perturbations caused by neighbouring planets and Jupiter. The orbit of Mercury is subject to such perturbations caused by Venus and Jupiter. Newtonian theory had accounted for all but one percent of the observed perihelion advance. It was this small discrepancy of 43 minutes of arc per century that general relativity was able to explain.

36 Kline (1987) p. 362.

37 Cited in Kline (1987) p. 524.

38 Skidelsky (1992a) p. 407.

39 Cited in Moore (1989) p. 373.

40 Cited in Moore (1989) p. 384.

41 Cited in Graham (1981) p. 128.

42 Heisenberg (1971) p. 68.

43 Heisenberg (1971) pp. 68–69. Einstein was not fully convinced: 'control by experiment is, of course, an essential prerequisite of the validity of any theory. But one can't possibly test everything. That is why I am so interested in your remarks about simplicity. Still, I should never claim that I really understood what is meant by the simplicity of natural laws'. See Heisenberg (1971) p. 69. Einstein raises an interesting point here. Experimentation is vital for theoretical development; but not every theory, nor even every part of a particular theory, is susceptible to testing. In such a case, appeal to aesthetic criteria is offered as a justification. However,

one scientist's simplicity is often another's complexity. The appeal to aesthetics is a subjective one, which runs against the grain of scientific objectivity. So we can understand Einstein's lament: 'what is meant by the simplicity of natural laws'.

44 Cited in Kanigel (1991) p. 7.
45 Cited in Davies (1985) p. 149.
46 Weinberg (1993) p. 119.
47 Weinberg (1993) p. 71.
48 Weinberg (1993) p. 131.
49 Weinberg (1993) p. 120.
50 Robert Oldershaw, 'What's wrong with the new physics?', *New Scientist*, 22 December 1990.
51 Weinberg and Salam were awarded the 1979 Nobel Prize for physics, together with the American Sheldon Glashow, who also made an important contribution to the field.
52 See Martinus J. G. Veltman (1986), 'The Higgs Boson', *Scientific American*, 255: 16.
53 James Peebles (1987), *Science*, 235: 372.
54 Cited in Davies and Brown (1988) p. 182.
55 Cited in Davies and Brown (1988) p. 194.
56 Steven Weinberg, 'Life in the Universe', *Scientific American*, October 1994.
57 In Davies and Brown (1988) pp. 182–184.
58 In Wolpert and Richards (1989) p. 47.
59 In Wolpert and Richards (1989) p. 48.
60 Cited in Chandrasekhar (1990) p. 69.
61 Cited in Chandrasekhar (1990) p. 60.
62 Cited in Chandrasekhar (1990) p. 60.
63 Hume (1975) pp. 291–2.
64 Barrow (1991) p. 183.
65 Casti (1992) p. 328.
66 Cited in Barrow (1992) p. 264.
67 Kline (1990) p. 1207.
68 Cited in Gregory (1988) p. 173.
69 Hardy (1941) pp. 123–4.
70 Ruelle (1991) pp. 12–13.
71 Einstein (1954) p. 233.
72 Feynman (1965) p. 57.

Bibliography

Adorno, T. and Horkheimer, M. (1989) *Dialectic of Enlightenment* (London: Verso).

Altvater, E. (1993) *The Future of the Market: An Essay on the Regulation of Money and Nature after the Collapse of 'Actually Existing Socialism'* (London: Verso).

Appleyard, B. (1989) *The Pleasures of the Peace: Art and Imagination in Post-War Britain* (London: Faber and Faber).

Appleyard, B. (1992) *Understanding the Present: Science and the Soul of Modern Man* (London: Picador).

Aronowitz, S. (1988) *Science as Power* (London: Macmillan).

Arrhenius, S. (1965) *Nobel Lectures in Physics* (Volume One) (New York: Elsevier).

Ayer, A. J. (1990) *Language, Truth and Logic* (London: Penguin).

Badash, L. (1969) *Rutherford and Boltwood* (New Haven: Yale University Press).

Baker, D. (1988), 'NASA Explores the High Frontier', *New Scientist*, 29 September.

Baker, G. L. and Gollub, J. P. (1990) *Chaotic Dynamics: An Introduction* (Cambridge: Cambridge University Press).

Barrow, J. D. and Tipler, F. J. (1989) *The Anthropic Cosmological Principle* (Oxford: Oxford University Press).

Barrow, J. D. (1990) *The World Within the World* (Oxford: Oxford University Press).

Barrow, J. D. (1991) *Theories of Everything: The Quest for Ultimate Explanation* (Oxford: Oxford University Press).

Barrow, J. D. (1992) *Pi in the Sky: Counting, Thinking, and Being* (Oxford: Clarendon Press).

Bass, T. A. (1990) *The Newtonian Casino* (London: Penguin).

Behrens, C. B. A. (1967) *The Ancien Regime* (London: Thames & Hudson).

Beiser, A. (1987) *Concepts of Modern Physics* (Fourth Edition) (New York: McGraw-Hill Book Company).

Bell, J. S. (1988) *Speakable and Unspeakable in Quantum Mechanics* (Cambridge: Cambridge University Press).

Berman, M. (1983) *All That is Solid Melts into Air: The Experience of Modernity* (London: Verso).

Bernal, J. D. (1969) *Science in History* (4 volumes) (London: Penguin).

Bernstein, J. (1991) *Quantum Profiles* (Princeton: Princeton University Press).

Bhaskar, R. (1989) *Reclaiming Reality: A Critical Introduction to Contemporary Philosophy* (London: Verso).

Biquard, P. (1965) *Frédéric Joliot-Curie: The Man and His Theories* (London: Souvenir Press Ltd).

Boas, M. (1970) *The Scientific Renaissance, 1450–1630* (London: Fontana).

Bohm, D. and Peat, F. D. (1989) *Science, Order, and Creativity* (London: Routledge).

Bohr, N. (1958) *Atomic Theory and Human Knowledge* (New York: John Wiley).

Born, M. (1971) *The Born-Einstein Letters* (London: Macmillian).

Born, M. (1978) *My Life: Recollections of a Nobel Laureate* (London: Taylor and Francis).

Bramwell, A. (1989) *Ecology in the Twentieth Century: A History* (New Haven: Yale University Press).

Bramwell, A. (1994) *The Fading of the Greens: The Decline of Environmental Politics in the West* (New Haven: Yale University Press).

Briggs, J. and Peat, F. D. (1989) *Turbulent Mirror* (New York: Harper and Row).

Brooke, J. H. (1991) *Science and Religion: Some Historical Perspectives* (Cambridge: Cambridge University Press).

Bukharin, N. (1971) *Science at the Cross Roads* (London: Frank Cass and Company Limited).

Burcham, W. E. (1963) *Nuclear Physics* (London: Longman).

Burke, J. (1985) *The Day the Universe Changed* (London: BBC).

Bush, V. (1946) *Endless Horizons* (Washington: Public Affairs Press).

Butterfield, H. (1951) *The Origins of Modern Science, 1300–1800* (London: G. Bell).

Capra, F. (1980) *The Tao of Physics* (New York: Bantam Press).

Carr, E. H. (1990) *What is History?* (London: Penguin).

Carson, R. (1982) *Silent Spring* (London: Penguin).

Cassidy, D. C. (1992) *Uncertainty: The Life and Science of Werner Heisenberg* (New York: W. H. Freeman and Company).

Casti, J. L. (1992) *Searching for Certainty: What Science Can Know about the Future* (London: Scribners).

Chandrasekhar, S. (1990) *Truth and Beauty* (Chicago: The University of Chicago Press).

Chant, C. and Fauvel, J. (1980) (editors) *From Darwin to Einstein: Historical Studies on Science and Belief* (Essex: Open University Press).

Clark, R. W. (1979) *Einstein: The Life and Times* (London: Hodder and Stoughton).

Cohen, B. I. (editor) (1958) *Isaac Newton's Papers and Letters on Natural Philosophy* (Cambridge, Mass: Harvard University Press).

Cohen, J. and Stewart, J. (1994) *The Collapse of Chaos: Discovering Simplicity in a Complex World* (London: Viking).

Coley, N. and Hall, V. (editors) (1980) *Darwin to Einstein: Primary Sources on Science and Belief* (Essex: Longman).

Colodny, R. G. (editor) (1972) *Paradigms and Paradoxes: The Philosophical Challenge of the Quantum Domain* (Pittsburgh: University of Pittsburgh Press).

Colodny, R. G. (editor) (1986) *From Quarks to Quasars: Philosophical Problems of Modern Physics* (Pittsburgh: University of Pittsburgh Press).

Collins, H. and Pinch, T. (1993) *The Golem: What Everyone Should Know about Science* (Cambridge: Cambridge University Press).

Corn, J. J. (editor) (1986) *Imagining Tomorrow: History, Technology, and the American Future* (Cambridge: MIT Press).

Cottingham, J. (1988) *The Rationalists* (Oxford: Oxford University Press).

Coveney, P. and Highfield, R. (1990) *The Arrow of Time* (London: W. H. Allen).

Crease, R. P. and Mann, C. C. (1986) *The Second Creation: Makers of the Revolution in Twentieth-Century Physics* (New York: Macmillan Publishing Company).

Crichton, M. (1991) *Jurassic Park* (London: Arrow).

Cunningham, A. and Jardine, N. (editors) (1990) *Romanticism and the Sciences* (Cambridge: Cambridge University Press).

Cushing, J. T. (1994) *Quantum Mechanics: Historical Contingency and the Copenhagen Hegemony* (Chicago: The University of Chicago Press).

Cushing, J. T. and McMullin, E. (editors) (1989) *Philosophical Consequences of Quantum Theory: Reflections on Bell's Theorem* (Notre Dame, Indiana: University of Notre Dame Press).

de Chardin, P. T. (1959) *The Phenomenon of Man* (London: Collins).

d'Espagnat, B. (1983) *In Search of Reality* (New York: Springer-Verlag).

d'Espagnat, B. (1989) *Reality and the Physicist* (Cambridge: Cambridge University Press).

de Vaudrey Heathcote, N. H. (1954) *Nobel Prize Winners in Physics 1901–1950* (New York: Schuman).

Darwin, C. (1985) *The Origin of Species: By Means of Natural Selection: Or the Preservation of Favoured Races in the Struggle for Life* (London: Penguin).

Davidson, J. D. and Rees-Mogg, W. (1992) *The Great Reckoning: How the World Will Change in the Depression of the 1990s* (London: Sidgwick and Jackson).

Davies, P. C. W. (1984) *Quantum Mechanics* (London: Routledge and Kegan Paul).

Davies, P. C. W. (1985) *Superforce* (London: Unwin Hyman).

Davies, P. C. W. (1987) *God and the New Physics* (London: Penguin).

Davies, P. C. W. (1989) *The Cosmic Blueprint* (London: Unwin).

Davies, P. C. W. (1993) *The Mind of God: Science and the Search for Ultimate Meaning* (London: Penguin Books).

Davies, P. C. W. and Brown, J. R. (1986) *The Ghost in the Atom* (Cambridge: Cambridge University Press).

Davies, P. C. W. and Brown, J. R. (editors) (1988) *Superstrings: A Theory of Everything?* (Cambridge: Cambridge University Press).

Davies, P. C. W. and Gribbin, J. (1991) *The Matter Myth: Towards Twenty-First-Century Science* (London: Viking).

Dawkins, R. (1988) *The Blind Watchmaker* (London: Penguin).

Dawkins, R. (1989) *The Extended Phenotype* (Oxford: Oxford University Press).

Dennett, D. C. (1992) *Consciousness Explained* (London: Viking).

Descartes, R. (1968) *Discourse on Method and the Meditations* (London: Penguin).

Devine, R. A. (1993) *The Sputnik Challenge* (Oxford: Oxford University Press).

Diamond, J. (1992) *The Rise and Fall of the Third Chimpanzee* (London: Vintage).

Dijksterhuis, E. J. (1961) *The Mechanization of the World Picture* (Oxford: Oxford University Press).

Dirac, P. A. M. (1963) 'The Evolution of the Physicists Picture of Nature', *Scientific American*, May 1963.

Duhem, P. (1977) *The Aim and Structure of Physical Theory* (New York: Atheneum).

Durant, J. (editor) (1992) *Public Understanding of Science,* 1: 1. (London: Science Museum Library).

Eagleton, T. (1991) *Ideology: An Introduction* (London: Verso).

Easlea, B. (1980) *Liberation and the Aims of Science: An Essay on the Obstacles to the Building of a Beautiful World* (Edinburgh: Scottish Academic Press).

Ehrlich, A. and P. (1987) *Earth* (London: Thames Methuen).

Einstein, A. (1954) *Ideas and Opinions* (New York: Crown).

Fauvel, J., Flood, R., Shortland, M., and Wilson, R. (editors) (1989) *Let Newton Be!* (Oxford: Oxford University Press).

Ferris, T. (1988) *Coming of Age in the Milky Way* (London: Bodley Head).

Feynman, R. (1965) *The Character of Physical Law* (London: BBC Publications).

Fine, A. (1986) *The Shaky Game: Einstein, Realism and the Quantum Theory* (Chicago: The University of Chicago Press).

Folse, H. J. (1985) *The Philosophy of Niels Bohr: The Framework of Complementarity* (Amsterdam: North-Holland).

Freud, S. (1985) *Collected Works 12* (London: Penguin).

Füredi, F. (1992) *Mythical Past, Elusive Future: History and Society in an Anxious Age* (London: Pluto Press).

Galilei, G. (1953) *Dialogue Concerning the Two Chief World Systems* (Berkeley: University of California Press).

Garment, S. (1991) *Scandal: The Crisis of Mistrust in American Politics* (New York: Random House).

Gell-Mann, M. (1994) *The Quark and the Jaguar: Adventures in the Simple and the Complex* (London: Little, Brown and Company).

Gerth, H. H. and Mills, C. W. (editors) (1977) *From Max Weber: Essays in Sociology* (London: Routledge and Kegan Paul).

Gibbins, P. (1987) *Particles and Paradoxes: The Limits of Quantum Logic* (Cambridge: Cambridge University Press).

Gill, P. (1986) *A Year in the Death of Africa* (London: Paladin Grafton Books).

Gillispie, C. C. (1990) *The Edge of Objectivity* (Princeton: Princeton University Press).

Gleick, J. (1988) *Chaos* (London: Sphere Books).

Goldsmith, M., Mackay, A., and Woudhuysen, J. (1980) *Einstein: The First 100 Years* (London: Pergamon Press).

Gore, A. (1992) *Earth in the Balance: Forging a New Common Purpose* (London: Earthscan Publications).

Gould, S. J. (1981) *The Mismeasure of Man* (London: Penguin).

Gould, S. J. (1990) *Wonderful Life: The Burgess Shale and the Nature of History* (London: Hutchinson Radius).

Gould, S. J. (1990) *The Individual in Darwin's World* (Edinburgh: Edinburgh University Press).

Graham, L. R. (1981) *Between Science and Values* (New York: Columbia University Press).

Gregory, B. (1988) *Inventing Reality* (New York: Wiley).

Gribbin, J. (1985) *In Search of Schrödinger's Cat* (London: Corgi Books, Transworld).

Gross, P. R. and Levitt, N. (1994) *Higher Superstition: The Academic Left and its Quarrels With Science* (Baltimore: The Johns Hopkins University Press).

Hall, N. (editor) (1992) *The New Scientist Guide to Chaos* (London: Penguin).

Hall, R. A. (1983) *The Revolution in Science 1500–1750* (London: Longman).

Hampson, N. (1968) *The Enlightenment* (London: Penguin).

Handy, C. (1994) *The Empty Raincoat: Making Sense of the Future* (London: Hutchinson).

Hankins, T. L. (1985) *Science and the Enlightenment* (Cambridge: Cambridge University Press).

Hardin, G. (1993) *Living Within Limits: Ecology, Economics, and Population Taboos* (New York: Oxford University Press).

Hardy, G. H. (1941) *A Mathematician's Apology* (Cambridge: Cambridge University Press).

Harris, E. E. (1991) *Cosmos and Anthropos: A Philosophical Interpretation of the Anthropic Cosmological Principle* (New Jersey: Humanities Press International).

Harrison, P. (1993) *The Third Revolution: Population, Environment and a Sustainable World* (London: Penguin Books).

Hayek, F. A. (1990) *The Fatal Conceit: The Errors of Socialism* (London: Routledge).

Hawking, S. W. (1988) *A Brief History of Time* (London: Bantam Press).

Heisenberg, W. (1971) *Physics and Beyond* (London: Allen & Unwin).

Hiley, B. J. and Peat, F. D. (editors) (1991) *Quantum Implications: Essays in Honour of David Bohm* (London: Routledge).

Hobsbawm, E. (1988a) *The Age of Revolution, Europe 1789–1848* (London: Cardinal Books).

Hobsbawm, E. (1988b) *The Age of Capital: 1848–1875* (London: Cardinal Books).

Hobsbawm, E. (1994) *Age of Extremes: The Short Twentieth Century 1914–1991* (London: Michael Joseph).

Holton, G. (1986) *The Advancement of Science, and its Burdens* (Cambridge: Cambridge University Press).

Holton, G. (1988) *Thematic Origins of Scientific Thought: Kepler to Einstein* (Cambridge, Massachusetts: Havard University Press).

Holton, G. (1993) *Science and Anti-Science* (Cambridge, Massachusetts: Harvard University Press).

Hooykaas, R. (1984) *Religion and the Rise of Modern Science* (Edingurgh: Scottish Academic Press).

Hoyer, U. (1981) *Niels Bohr: Collected Works, Volume 2, Work on Atomic Physics (1912–17)* (Amsterdam: North Holland).

Hubbard, R. and Wald, E. (1993) *Exploding the Gene Myth* (Boston, Massachusetts: Beacon Press).

Hughes, H. S. (1979) *The Reorientation of European Social Thought 1890–1930* (Brighton: Harvester Press).

Hume, D. (1975) *Enquiries Concerning Human Understanding and Concerning the Principles of Morals* (Oxford: Oxford University Press).

Hurt, H. (1988) *For All Mankind* (London: Queen Anne Press).

Huxley, A. (1977) *Brave New World* (London: Grafton Books).

Jammer, M. (1966) *The Conceptual Development of Quantum Mechanics* (New York: McGraw-Hill).

Jantsch, E. (1992) *The Self Organising Universe: Scientific and Human Implications of the Emerging Paradigm of Evolution* (Oxford: Pergamon).

Jauch, J. M. (1989) *Are Quanta Real?* (Bloomington: Indiana University Press).

Jennings, H. (1985) *Pandaemonium: The Coming of the Machine as Seen by Contemporary Observers* (London: Andre Deutsch Ltd).

Jones, G. (1988) *Science, Politics and the Cold War* (London: Routledge).

Jungk, R. (1958) *Brighter than a Thousand Suns* (London: Victor Gollancz).

Kanigel, R. (1991) *The Man Who Knew Infinity: A Life of the Genius Ramanujan* (London: Abacus).

Kellert, S. H. (1994) *In the Wake of Chaos* (Chicago: The University of Chicago Press).

Kennedy, P. (1994) *Preparing for the Twenty-First Century* (London: Fontana Press).

Kern, S. (1983) *The Culture of Time and Space: 1880–1918* (Cambridge, Massachusetts: Harvard University Press).

Kevles, D. J. (1985) *In the Name of Eugenics: Genetics and the Uses of Human Heredity* (New York: Knopf).

Kevles, D. J. (1987) *The Physicists: The History of a Scientific Community in Modern America* (Cambridge, Massachusetts: Harvard University Press).

Kevles, D. J. and Hood, L. (editors) (1992) *The Code of Codes: Scientific and Social Issues in the Human Genome Project* (Cambridge, Massachusetts: Harvard University Press).

Kiernan, C. (1973) *The Enlightenment and Science in Eighteenth-Century France* (Banbury, Oxfordshire: The Voltaire Foundation).

Kitchener, R. F. (1988) *The World View of Contemporary Physics: Does it Need a New Metaphysics?* (Albany, New York: State University of New York Press).

Kline, M. (1980) *Mathematics and the Loss of Certainty* (Oxford: Oxford University Press).

Kline, M. (1987) *Mathematics in Western Culture* (Oxford: Oxford University Press).

Kline, M. (1990) *Mathematical Thought: From Ancient to Modern Times* (Oxford: Oxford University Press).

Koestler, A. (1972) *The Roots of Coincidence* (London: Hutchinson).

Koestler, A. (1986) *The Sleepwalkers: A History of Man's Changing Vision of the Universe* (London: Penguin Books).

Komesaroff, P. (1986) *Objectivity, Science and Society* (London: Routledge and Kegan Paul).

Kuhn, A. (editor) (1990) *Alien Zone: Cultural Theory and Contemporary Science Fiction Cinema* (London: Verso).

Kuhn, T. S. (1970) *The Structure of Scientific Revolutions* (Chicago: The University of Chicago Press).

LaFollatte, M. C. (1990) *Making Science Our Own* (Chicago: The University of Chicago Press).

Lakatos, I. and Musgrave, A. (editors) (1970) *Criticism and the Growth of Knowledge* (Cambridge: Cambridge University Press).

Landes, D. S. (1989) *The Unbound Prometheus* (Cambridge: Cambridge University Press).

Laurikainen, K. V. (1985) *Beyond the Atom: The Philosophical Thought of Wolfgang Pauli* (Berlin: Springer-Verlag).

Layzer, D. (1990) *Cosmogenesis: The Growth of Order in the Universe* (Oxford: Oxford University Press).

Lederman, L. (1991) *Science: The End of the Frontier?* (Washington, DC: American Association for the Advancement of Sciences).

Levy, S. (1992) *Artificial Life: The Quest for a New Creation* (London: Jonathan Cape).

Lewontin, R. (1993) *The Doctrine of DNA* (London: Penguin Books).

Lindberg, D. C. and Westman, R. S. (editors) (1990) *Reappraisals of the Scientific Revolution* (Cambridge: Cambridge University Press).

Lock, F. P. (1985) *Burke's Reflections on the Revolution in France* (London: George Allen and Unwin).

Locke, J. (1990) *An Essay Concerning Human Understanding* (Oxford: Clarendon Press).

Lorenz, E. L. (1993) *The Essence of Chaos* (London: UCL Press Limited).

Lovelock, J. (1988) *Gaia – A New Look at Life on Earth* (Oxford: Oxford University Press).

Lovelock, J. (1989) *The Ages of Gaia* (Oxford: Oxford University Press).

Lucas, G. R. (Jnr) (1989) *The Rehabilitation of Whitehead: An Analytic and Historical Assesment of Process Philosophy* (New York: State University of New York Press).

Lukács, G. (1980) *The Destruction of Reason* (London: The Merlin Press).

Marks, J. (1983) *Science and the Making of the Modern World* (London: Heinemann).

Maier, C. S. (1987) *In Search of Stability: Explorations in Historical Political Economy* (Cambridge: Cambridge University Press).

Manchester, W. (1992) *A World Lit Only by Fire: The Medieval Mind and the Renaissance* (Boston: Little, Brown and Company).

Marcuse, H. (1964) *One Dimensional Man: Studies in the Ideology of Advanced Industrial Society* (London: Routledge, Keegan-Paul).

Mason, S. F. (1962) *A History of the Sciences* (New York: Collier Books).
McClelland, J. S. (1970) *The French Right: From de Maistre to Maurras* (London: Jonathan Cape).
McCormmach, R. (1970), 'The First Phase of the Bohr–Einstein Dialogue', *Historical Studies in the Physical Sciences:* 2, 1.
McDougall, W. A. (1985) *The Heavens and the Earth: A Political History of the Space Age* (Basic Books, New York).
McGuinnes, B. (editor) (1974) *Ludwig Boltzman: Theoretical Physics and Philosophical Problems: Selected Writings* (Dordrecht-Holland: D. Reidel Publishing Company).
McKibben, B. (1990) *The End of Nature* (London: Viking).
Mehra, J. and Rechenberg, H. (1982) *Historical Development of Quantum Theory, Volume 1* (New York: Springer-Verlag).
Mermin, N. D. (1990) *Boojums All the Way Through: Communicating Science in a Prosaic Age* (Cambridge: Cambridge University Press).
Mészáros, I. (1989) *The Power of Ideology* (London: Harvester Wheatsheaf).
Miliband, R. and Panitch, L. (1993) *Socialist Register 1993: Real Problems False Solutions* (London: The Merlin Press).
Monk, R. (1991) *Ludwig Wittgenstein: The Duty of Genius* (London: Vintage).
Monod, J. (1972) *Chance and Necessity* (London: Collins).
Moore, W. (1989) *Schrödinger: Life and Thought* (Cambridge: Cambridge University Press).
Morris, R. (1992) *The Edges of Science: Crossing the Boundary from Physics to Metaphysics* (London: Fourth Estate).
Murdoch, D. (1989) *Niels Bohr's Philosophy of Physics* (Cambridge: Cambridge University Press).
Newman, J. R. (1947), 'America's Most Radical Law', *Harper's Monthly Magazine,* May 1947.
Newton-Smith, W. H. (1990) *The Rationality of Science* (London: Routledge).
Nicolis, G. and Prigogine, I. (1989) *Exploring Complexity* (New York: W. H. Freeman and Company).
Nye, R. A. (1975) *The Origins of Crowd Psychology: Gustave LeBon and the Crisis of Mass Democracy in the Third Republic* (London: Sage Publications).
Omnès, R. (1994) *The Interpretation of Quantum Mechanics* (Princeton, New Jersey: Princeton University Press).
O'Neill, J. J., 'Enter Atomic Power', *Harper's Monthly Magazine,* June 1940.
Pagels, H. R. (1984) *The Cosmic Code: Quantum Physics as the Language of Nature* (Harmondsworth: Penguin).
Pais, A. (1982) *Subtle is the Lord... : The Science and the Life of Albert Einstein* (Oxford: Clarendon Press).
Pais, A. (1991) *Niels Bohr's Times: In Physics, Philosophy, and Polity* (Oxford: Clarendon Press).
Pais, A. (1994) *Einstein Lived Here* (Oxford: Oxford University Press).
Parker, D. and Stacey, R. (1994) *Chaos, Management and Economics: The Implications of Non-Linear Thinking* (London: Institute of Economic Affairs).
Peat, F. D. (1990) *Einstein's Moon: Bell's Theorem and the Curious Quest for Quantum Reality* (Chicago: Contemporary Books).

Penrose, R. (1989) *The Emperor's New Mind* (Oxford: Oxford University Press).

Peters, E. E. (1992) *Chaos and Order in the Capital markets: A New View of Cycles, Prices, and Market Volatility* (New York: John Wiley).

Pick, D. (1989) *Faces of Degeneration: A European Disorder, c1848–c1918* (Cambridge: Cambridge University Press).

Poincaré, H. (1905) *Science and Method* (New York: Dover).

Polanyi, P. (1969) *Personal Knowledge: Towards a Post-Critical Philosophy* (London: Routledge and Kegan Paul).

Popper, K. (1968) *The Logic of Scientific Discovery* (London: Hutchinson).

Popper, K. (1986) *The Poverty of Historicism* (London: ARK Paperbacks).

Popper, K. (1992) *Unended Quest: An Intellectual Autobiography* (London: Routledge).

Poudstone, W. (1993) *Prisoner's Dilemma: John von Neumann, Game Theory, and the Puzzle of the Bomb* (Oxford: Oxford University Press).

Powers, T. (1993) *Heisenberg's War: The Secret History of the German Bomb* (London: Jonathan Cape).

Prigogine, I. and Stengers, I. (1984) *Order Out of Chaos* (London: Fontana).

Quine, W. V. O. (1970) *The Web of Belief* (New York: Random House).

Rae, A. I. M. (1986) *Quantum Physics: Illusion or Reality?* (Cambridge: Cambridge University Press).

Rassam, C. C. (1993) *The Second Culture: British Science in Crisis – the Scientists Speak Out* (London: Aurum Press).

Redhead, M. L. G. (1989) *Incompleteness, Nonlocality and Realism* (Oxford: Clarendon Press).

Reeves, H. (1991) *The Hour of Our Delight: Cosmic Evolution, Order and Complexity* (New York: W. H. Freeman and Company).

Rhodes, R. (1986) *The Making of the Atomic Bomb* (London: Simon and Schuster).

Rich, B. (1994) *Mortgaging the Earth: The World Bank, Environmental Impoverishment and the Crisis of Development* (London: Earthscan Publications Limited).

Robertson, J. M. (editor) (1905) *The Philosophical Works of Francis Bacon* (London: Routledge).

Rose, S. (1992) *The Making of Memory: From Molecules to Mind* (London: Bantam Press).

Ross, A. (1991) *Strange Weather: Culture, Science, and Technology in the Age of Limits* (New York: Verso).

Roszak, T. (1971) *The Making of a Counter Culture: Reflections on the Technocratic Society and its Youthful Opposition* (London: Faber and Faber).

Rubin, I. I. (1979) *A History of Economic Thought* (London: Ink Links).

Ruelle, D. (1991) *Chance and Chaos* (Princeton: Princeton University Press).

Russell, B. (1985) *The Impact of Science on Society* (London: Unwin Hyman).

Russell, B. (1990) *A History of Western Philosophy* (London: Unwin).

Russell, C. (1983) *Science and Social Change 1700–1900* (London: Macmillan).

Schmidt, A. (1971) *The Concept of Nature in Marx* (London: New Left Books).

Schrödinger, E. (1961) *Science and Humanism* (Cambridge: Cambridge University Press).

Schrödinger, E. (1992) *What is life?* (Cambridge: Cambridge University Press).

Schumacher, E. F. (1988) *Small is Beautiful* (London: Abacus).

Scott, P. (1990) *Knowledge and Nation* (Edinburgh: Edinburgh University Press).

Segrè, E. (1980) *From X-rays to Quarks: Modern Physicists and Their Discoveries* (San Francisco: W. H. Freeman and Company).

Segrè, E. (1984) *From Falling Bodies to Radio Waves: Classical Physicists and Their Discoveries* (New York: W. H. Freeman and Company).

Sheldrake, R. (1989) *The Presence of the Past* (London: Fontana).

Sheldrake, R. (1990) *The Rebirth of Nature: The Greening of Science and God* (London: Century).

Shelley, M. (1985) *Frankenstein* (London: Penguin).

Skidelsky, R. (1992a) *John Maynard Keynes: Hopes Betrayed 1883–1920* (London: Macmillan).

Skidelsky, R. (1992b) *John Maynard Keynes: The Economist as Saviour 1920–1937* (London: Macmillan).

Smith, J. A. (1991) *The Idea Brokers: Think Tanks and the Rise of the New Policy Élite* (New York: The Free Press).

Snow, C. P. (1993) *The Two Cultures* (Cambridge: Cambridge University Press).

Snow, C. P. (1981) *The Physicists* (London: Macmillan).

Soffer, R. N. (1978) *Ethics and Society in England: The Revolution in the Social Sciences 1870–1914* (Berkeley: University of California Press).

Solomon, R. C. (1990) Continental Philosophy *Since 1750: The Rise and Fall of the Self* (Oxford: Oxford University Press).

Spadafora, D. (1990) *The Idea of Progress in Eighteeth Century Britain* (New Haven: Yale University Press).

Speziali, P. (1972) (editor) *Albert Einstein-Michele Besso Correspondence 1903–1955* (Paris: Hermann).

Stewart, I. (1989) *Does God Play Dice?* (London: Basil Blackwell).

Stuewer, R. H. (1975) *The Compton Effect: Turning Point in Physics* (New York: Science History Publications).

Therborn, G. (1980) *Science Class and Society* (London: Verso).

Tilby, A. (1992) *SOUL: An Introduction to the New Cosmology – Time, Consciousness and God* (London: BBC Education).

Toulmin, S. (1990) *Cosmopolis: The Hidden Agenda of Modernity* (Chicago: The University of Chicago Press).

Trigg, R. (1989) *Reality at Risk: A Defence of Realism in Philosophy and the Sciences* (London: Harvester Wheatsheaf).

van Fraassen, B. C. (1991) *Quantum Mechanics: An Empiricist View* (Oxford: Clarendon Press).

Waldrop, M. M. (1992) *Complexity: The Emerging Science at the Edge of Order and Chaos* (London: Viking).

Walker, M. (1989) *German National Socialism and the Quest for Nuclear Power: 1939–1949* (Cambridge: Cambridge University Press).

Weber, R. (1981) *Pioneers of Science: Nobel Prize Winners in Physics* (London: Scientific Book Club).

Weber, R. (1990) *Dialogues With Scientists and Sages* (London: Penguin).

Weinberg, S. (1993) *Dreams of a Final Theory* (London: Hutchinson Radius).

Weisskopf, V. F. (1989) *The Privilege of Being a Physicist* (New York: W. H. Freeman and Company).

Westfall, R. (1988) *The Construction of Modern Science: Mechanisms and Mechanics* (Cambridge: Cambridge University Press).

Wheeler, J. A. and Zurek, W. H. (editors) (1983) *Quantum Theory and Measurement* (Princeton: Princeton University Press).

Whitehead, A. N. (1985) *Science and the Modern World* (London: Free Association Books).

Wigner, E. P. (1967) *Symmetries and Reflections* (Bloomington: Indiana University Press).

Wilkie, T. (1991) *British Science and Politics Since 1945* (Oxford: Blackwell).

Wilkie, T. (1993) *Perilous Knowledge: The Human Genome Project and its Implications* (London: Faber and Faber).

Wittgenstein, L. (1988) *Tractatus Logico-Philosophicus* (London: Routledge).

Wolpert, L. and Richards, A. (editors) (1989) *A Passion for Science* (Oxford: Oxford University Press).

Wolpert, L. (1992) *The Unnatural Nature of Science* (London: Faber and Faber).

Zeitlin, I. (1981) *Ideology and the Development of Sociological Theory* (New Jersey: Prentice Hall).

Zohar, D. (1990) *The Quantum Self: A Revolutionary View of Human Nature and Consciousness in the New Physics* (London: Bloomsbury).

Zukav, G. (1980) *The Dancing Wu Li Masters* (New York: Bantam Press).

Zurek, W. H. (1990) *Complexity, Entropy and the Physics of Information* (Redwood City, California: Addison Wesley).

INDEX